AF839333

How Words Help Us Think

Also available from Bloomsbury:

The Philosophy and Science of Predictive Processing edited by Dina Mendonça, Manuel Curado and Steven S. Gouveia
Advances in Experimental Philosophy of Mind edited by Justin Sytsma
Philosophy in a Technological World by James Tartaglia

How Words Help Us Think

An Externalist Account of Representational Intentionality

Nancy Salay

BLOOMSBURY ACADEMIC
LONDON • NEW YORK • OXFORD • NEW DELHI • SYDNEY

BLOOMSBURY ACADEMIC
Bloomsbury Publishing Plc
50 Bedford Square, London, WC1B 3DP, UK
1385 Broadway, New York, NY 10018, USA
29 Earlsfort Terrace, Dublin 2, Ireland

BLOOMSBURY, BLOOMSBURY ACADEMIC and the Diana logo are trademarks of Bloomsbury Publishing Plc

First published in Great Britain 2025

Copyright © Nancy Salay 2025

Nancy Salay has asserted her right under the Copyright, Designs and Patents Act, 1988, to be identified as Author of this work.

For legal purposes the Acknowledgments on p. viii constitute an extension of this copyright page.

Cover image: francescoch / iStock

All rights reserved. No part of this publication may be reproduced or transmitted in any form or by any means, electronic or mechanical, including photocopying, recording, or any information storage or retrieval system, without prior permission in writing from the publishers.

Bloomsbury Publishing Plc does not have any control over, or responsibility for, any third-party websites referred to or in this book. All internet addresses given in this book were correct at the time of going to press. The author and publisher regret any inconvenience caused if addresses have changed or sites have ceased to exist, but can accept no responsibility for any such changes.

A catalogue record for this book is available from the British Library.

A catalog record for this book is available from the Library of Congress.

ISBN: HB: 978-1-3502-6682-7
ePDF: 978-1-3502-6683-4
eBook: 978-1-3502-6684-1

Typeset by Deanta Global Publishing Services, Chennai, India

To find out more about our authors and books visit www.bloomsbury.com and sign up for our newsletters.

Dedicated to the paradigm shifters who have inspired me

Contents

Acknowledgments

This book is the culmination of thoughts that have been percolating over the past ten years. I owe a debt of inspiration to thinkers such as Andy Clark, William Bechtel, Randall Beer, Anthony Chemero, Carl Craver, Daniel Dennett, Hubert Dreyfus, Walter Freeman, Stephen Gallagher, Fred Keijzer, Mary Midgley, Alva Noë, Hilary Putnam, Eleanor Rosch, Esther Thelen, Evan Thompson, Francisco Varela, and Michael Wheeler (University of Stirling), who are notable, to me, for offering new ways of thinking about old problems. In various ways, they have motivated the ideas in this book.

Friends and colleagues who have helped me think through these ideas, sometimes unknowingly, also deserve special thanks: Selim Akl, David Bakhurst, Cheryl Breukelman, Sue Donaldson, Paul Fairfield, Meredith Chivers, Tom Hollenstein, Will Kymlicka, Henry Laycock, Josh Mozersky, Elliot Paul, and Catherine Stinton. Thank you also to Pietro Perconti for inviting me to share some of these ideas at the Science of Consciousness Congress 2023 in lovely Taormina—the many comments and discussions afterward were helpful.

The externalist analysis of AI in Chapter 12 was much improved by what I learned from the intrepid folk who made the inaugural session of PRAID (Workshop on Perception and Artificial Intelligence) such a success: Gunnar Blohm, Jeff Bowers, Cristina Conati, Sidney Givigi, Ting Hu, Donald Landes, Matthew Pan, Susanna Schellenberg, Katharina Schwarz, Michael Wheeler (Queen's University), and most especially Monica Castelhano, who cohosted with me and without whose help it would not have happened, and Hannah Lum Smith, for organizing us all.

Thank you to the lovely people at Bloomsbury Press—Aimee Brown, Colleen Coalter, and Suzie Nash—who have cheerfully kept this project on track. They have been a pleasure to work with. Thank you also to an anonymous referee, whose comments have been so helpful. Much appreciation and thanks go to

Raymond Russ, the editor-in-chief at *Mind and Behavior*, who gave permission for including some of the material that originally appeared in:

Salay, N. (2019), "Learning How to Represent: An Associationist Account," *The Journal of Mind and Behavior*, 40 (2): 121–145.

Salay, N. (2021), "How Reductive Analyses of Content Are Confused and How to Fix Them: A Critique of Caritel Semantics," *The Journal of Mind and Behavior*, 42 (2): 109–138.

Finally, thank you to my family, without whose support I could not have completed what seemed, at times, a Sisyphean task: to my brother, Rick, who went the extra mile and helped me format references so that I got my first draft in on time, thank you; to my parents, for raising me with so much love, and for encouraging me to follow my dreams, thank you; to my wise-beyond-his-years son, Griffin, who told me precisely what I needed to hear whenever I felt discouraged, thank you; and to my husband, Morgan Wade, for his unflagging love and support in both word and deed, thank you, thank you, thank you.

This book was funded, in part, by the Queen's University Fund for Scholarly Research and Creative Work and Professional Development.

Introduction

Consider this situation: Fred tramples through Sally's vegetable bed, destroying much of it.[1] When Sally angrily confronts Fred, he remorsefully explains that it was an accident: he was sleepwalking during the night, and this is something he has no conscious control over. Sally, surprised, softens. When Fred promises to replace what was damaged and help replant the garden, her anger abates.

Something odd is going on in this exchange. The physical features of the situation have not changed—the garden is still a mess—but Sally has gone from angry, certain that Fred's destructive behavior was thoughtless carelessness, to appeased and, perhaps, concerned for Fred. Why? We might say that Fred's words have painted a picture for Sally of what actually happened during the night (unintentional damage to her garden) and what *might* unfold (putting her garden to right). Sally's original anger is fueled by the current situation, the ruined garden, but her anger dissipates when she takes this wider view, seeing not only what is here and now, but what has passed and what might come. Two questions arise.

(1) How can Sally entertain and respond to what is not actually in front of her?
(2) How have Fred's words made these alternatives present for her?

The first question has traditionally been taken as part of the Problem of Intentionality:[2] that human beings can plan for the future and regret the past, react to what a sentence or picture represents, in short, think and talk *about* the stuff of our world, seems clear, but how we do this, in what this capacity consists, is still unknown. The second question has been treated as secondary to this problem: How do intentional beings use language to *communicate* their internal thoughts to one another? When Fred exclaims, "I will help you fix this, I promise!" Sally doesn't say, "you don't have to yell, I'm right here!" She responds to the future they represent, and her stance softens. On this model, the ability to use language is dependent on a more fundamental representational ability.

The conceit of this book is that the reverse is true: our capacity to deliberate—to consciously think *about* things—is a consequence of our linguistic prowess, not a precondition of it. On this view, the key contribution of language is the cognitive skill to represent it affords, not communication.

Getting this right is important. Deliberative reasoning is the root of our rationality. As rational beings, we aren't (just) caused to behave in certain ways, as a piece of paper floating this way and that upon the wind is: we do things for reasons. That we act on our choices is, in turn, the ground of the normative systems that govern our familial, social, legal, and political lives: people are either blamable or praiseworthy for their actions *because* they choose to perform them. Where there is no intention[3] to behave in a certain way, culpability falls away. Sally is placated by Fred's explanation because she realizes that Fred did not mean to ruin her garden; it was an accident. If our best cognitive theory cannot distinguish between actions that are performed for reasons and those that are merely causal consequences of local physical factors, then we will have no sound basis on which to make normative judgments at all. As increasingly sophisticated AI systems enter the public space, interacting with us, and replacing us in our jobs, very soon we will need to make normative judgments about how and to what extent they are responsible for *their* behavior. Unless we have a theory of reasoning to guide these judgments, we will be woefully unprepared to do so.

Unfortunately, we are still lacking such an account. Some are on offer[4] but, as I argue in Chapter 6, internalist approaches alone will not suffice. On those accounts, our deliberative capacities are grounded in a more fundamental representational one: "Human survival critically depends on the ability to anticipate and prepare for likely future events in the absence of sensory stimulation (Bar 2007; Holland 1990; Ingvar 1985). Central to this capacity is *the ability to recur on mental representations of prior experiences*" [emphasis added] (Diekhof et al. 2011: 1703). Such representationalism lies behind the bulk of research in cognitive science today.

A competing approach argues that our cognitive capacities are the result of the complex coordination of lower-level processes with embodiment and environmental factors, not representational relations: "When we examine very simple level intelligence we find that explicit representations and models of the world simply get in the way. It turns out to be better to use the world as its own model" (Brooks 1991A: 139).

Each of these views has insights we must keep, but the debate itself, focused as it is on whether representations are involved in deliberative activity, is orthogonal to the real locus of the issue, namely, whether deliberative activity is reducible to brain activity. As I will argue, the answer is no.[5] The externalist view being developed here picks up some of the provocative suggestions of Extended Mind Theory (Clark and Chalmers 1998) on which our cognitive capacities are partly constituted by the external scaffolding, "widewear" (Clark 1997), that supports them. We develop algorithms that extend our computational capacities, long division for example, and we off-load memory tasks onto our environment: a string around the finger reminds one to pick up the dry-cleaning; a to-do list keeps track of tasks; and Alzheimer's patients achieve independence with the aid of sticky notes. Language itself, Clark (2002, 2006a, b) and others (Cimatti and Vallortigara 2015; Deacon 1998; Donald 1993; Logan 2007; Wheeler 2004, 2005, 2019) have argued, is a cognitive niche construction, a tool that "transforms problem spaces in ways that aid thinking and reasoning" (Clark 2006a, b). On this view, the logic-like structure of language is a cognitive resource that complements the pattern-completing capacities of our brains, allowing us to formulate inferences, make plans, think about tasks in a linear fashion, and so on, all cognitive activities that a human without language could not accomplish.

Externalism, then, finds a middle ground in the representation/anti-representation debate: representations do have a powerful role to play in deliberative processes, not as internal structures that we "recur" on, but as external tools for spatiotemporally extending situations. On this view, a deliberative capacity, what I will be calling "representational intentionality," is a strongly scaffolded skill rather than a fundamental capacity. Externalism holds that while neural activity has a critical role to play in all cognitive behavior, the development of representational intentionality requires a particular sort of skill and a certain kind of environment. Elaborating on these central ideas will require some stage-setting.

Part I functions as a backgrounder for the central work of Part II. In Chapter 1, I set out the Problem of Representational Intentionality in detail and contextualize it within the two key philosophical traditions that have addressed it: Analytic Philosophy and Phenomenology. In Chapters 2–4,

I describe the dynamic mechanistic framework within which this account is developed and set out some key concepts.

Part II contains the body of the account. In Chapter 5, I revisit the Problem of Intentionality, breaking it down into three subproblems: the problems of Action Intentionality, Directed Intentionality, and Representational Intentionality, respectively. In Chapter 6, I explain why internalist approaches to the Problem of Representational Intentionality will not shed light on the question. In Chapters 7 and 8, I present the externalist solution.

Part III offers some implications of Externalism and, thereby, some avenues for testing its various hypotheses. There is much more work to be done on this front, but it is a start. In Chapter 10, I examine some empirical implications of the cognitive mechanism of expectation posited by Externalism. In Chapter 11, I give an externalist analysis of the cognitive capacities and limitations of our most promising AI systems. Finally, in Chapter 12, I engage with some of the philosophical implications of the view.

Part I

Setting the Framework

1

The Puzzle

Intentionality

Hamlet's admonishment, "There are more things in heaven and earth, Horatio, than are dreamt of in your philosophy" (Shakespeare 1.5.166), might be the tagline for a book on consciousness. We take for granted that we live in a world filled with other sentient agents, beings who notice us and who are aware of what we do. But when we stop to wonder about this sentient capacity, we are mystified. Conscious awareness seems importantly different from other kinds of awareness. My car emits a loud "beep" when the sensors arrayed along its rear bumper come too close to an object: the obstacle triggers the sensors which in turn trigger a mechanism for emitting a beep. In some sense, the car is aware of the obstacle. I am equipped with a similar sensory mechanism. When the sound waves produced by the beep hit my eardrum, sensory neurons are activated and these in turn trigger other neural activity in a cascade of interactions and I too become aware of the obstacle. But whereas the car is not *consciously* aware of the obstacle or its own beeping behavior, I am: as I apply the brakes, I might wonder, "Did the garbage can roll onto the driveway?" Explaining how my awareness is different from the car's is, as David Chalmers (1995) infamously observed, "the Hard Problem" (200). Fortunately, we do not need to answer it to make headway on the Problem of Intentionality. Indeed, progress with the latter will make the Hard Problem less hard because we will be better equipped to distinguish between conscious and intentional phenomena, something that is currently confused.

From the standpoint of intentionality, the salient difference between the car's behavior and mine is that the car is implementing a subroutine that is triggered by the physical presence of an object, while I am responding to what the car's beeping represents, namely that I am about to hit something. Furthermore, while the object on the driveway is not a *thing* for the car, for me it is a garbage can, mine. Whether we are consciously aware of it or not, our world is filled with objects and situations that we recognize, that mean something to us. Even when we move through it on autopilot, our world is filled with trees, people, cats, cars, blue skies, and garbage bins. As Alva Noë (2021) puts it: "The world is present to us, in thought, and in experience" (957). The car, on the other hand, has no world; there is nothing at all present for it. Intentional systems, then, are those for whom there is a world. There are different ways to understand what this means, as we shall see in a moment, but before we look at them, I will introduce some examples to ground the discussion to follow.

Recall the scenario in our opening chapter. When Fred tramples Sally's vegetable garden, Fred is regretful about the situation; he knows that Sally put time and effort into getting it to its current state. Sally, on the other hand, is angry about it; her time and effort feel wasted. These responses reflect, in a way yet to be explained, Fred's and Sally's respective intentional responses to the situation.

Some,[1] as we've already seen, analyze these intentional stances in terms of representation: Fred's regret about the situation is a relation between internal processes that represent the situation itself and others that represent Fred's attitudes toward it. Of course, simply citing representations cannot account for an intentional capacity since many systems have representational states, at least we like to describe them as such, and only some of those systems can think about their world. The bi-metallic strip inside a thermostat contracts and expands as the air temperature drops and rises and this, in turn, opens or closes a switch that controls the furnace. In being thus causally linked to the ambient temperature of the air, the thermostat's states represent it; indeed, we could hook up a digital readout to display this information. But none of the thermostat's activity is caused by the *information* its states represent. The thermostat triggers the furnace to turn on because the rise in temperature causes the strip to expand and trigger the power switch, not *because the temperature is 20°C*. Citing the information that the processes implicitly represent adds nothing to the account. In other words, representations have no explanatory force in such contexts.

Is the same true of Sally and Fred? Is their behavior a more complex series of physical causal connections that merely march in step with "mental" representations? Or is there something different going on in the intentional case? As will become clear through the course of this book, this is itself an ill-conceived question because it rests on the confused idea that Fred and Sally's intentional responses are representations at all, at least representations that play a role in their deliberations. Consider how their exchange might continue.

Fred is regretful and because of this regret he offers Sally a promise. He uses words to do this. He says, "I will pay for these damages." What sort of thing is this promise, and how do Fred's words create it? Here, I shall argue, we are warranted in invoking representations to explain what is happening. The promise is a picture, a representation of a future situation in which Sally's garden is repaired through Fred's efforts. The words that Fred uses—and they could have been different words—cause Sally to "see" this possible future too.

Fred and Sally seem to be doing something more here than intentionally responding to an occurrent situation: they are creating new situations by trading representations. Fred offers a picture, and Sally picks it up. It is to this expanded situation, one that includes the possible future Fred presents, that they intentionally respond. Sally and Fred, being intentional systems, are always already in a world filled with objects that mean something to them; in addition, they can use some of these objects to represent the world as it might be, to themselves and to one another.

Not all intentional systems can use representations in this way: a lion stalking its prey catches a scent on the wind as it changes direction; as it slinks to a new location, its senses are extended, continually checking to ensure that it remains downwind of its target. The lion is drawn to change position because it senses a shift in the wind and because of its desire for food, but these reasons are likely not explicitly represented for the lion; they work behind or beneath the scene, as it were. Fred and Sally, in contrast, can act for explicit reasons: "I will help you fix this because I feel just awful about what has happened." Fred and Sally can respond directly to what these words represent while the lion can respond only indirectly to what a situation means for it, by being tuned to it. Reasons thus sometimes play a direct causal role in the behavior of systems such as Fred and Sally, while they play only an indirect one in the behavior of systems such as the lion; and they play no role at all in the behavior of systems such as the thermostat. Or at least that's how it seems on this initial look at what might turn out to be only superficially distinct examples of intentional behavior. Whether we take the lion's behavior to be of the same sort as Sally's and Fred's, an account of intentionality will need to explain how they are either the same or different and, further, it will need to distinguish their behavior from the non-intentional sort exemplified by thermostats and the like.

Franz Brentano's (1874) specification of the Problem of Intentionality has grounded subsequent investigations so it will help to situate the two central approaches I will be considering in terms of it. Focusing on the mental aspect of intentionality, he notes:

> Every mental phenomenon is characterized by . . . the intentional (or mental)† inexistence of an object. . . . Every mental phenomenon includes something as object within itself, although they do not all do so in the same way. In presentation something is presented, in judgement something is affirmed or denied, in love loved, in hate hated, in desire desired and so on. This intentional in-existence is characteristic exclusively of mental phenomena. No physical phenomenon exhibits anything like it. (88)

One way to understand this specification is in terms of the concept of representation: to "include[s] something as object within itself" is to represent that object. We've considered this briefly already. On this view, Fred's regret and Sally's anger are responses to their representations of the situation: Fred thinks about the situation and is saddened; Sally thinks about it and becomes angry. Their respective thoughts represent—contain, in this sense—the situation itself. The lion's awareness of the change in wind direction can also be construed as a representation: the lion's perception of the scene is a picture of what is out there. Intentionality, on this view, is this representational capacity, of constructing internal pictures of the world that guide behavior.

Proponents of this approach tend to view perception as the most basic representational intentional capacity—to perceive X is to represent X—but while all animals have this capacity to some degree, fewer have the capacity to explicitly guide their behavior with representations, as we've already noted. Unlike lions, Sally and Fred can argue about their respective perceptions: for Sally, the purple door is "purple like red wine" while for Fred, it is more "mauve, like my favourite shirt." As representation-using agents in this sense, we make plans for the future, we have regrets about the past, and we make judgments about the present. The Problem of Intentionality, through this lens, is the problem of explaining in a non-question-begging way what this representation relation consists in, how it is distinct from other, non-representational, relations, and what makes one representation distinct from another.

Another way to understand Brentano's specification, however, is in terms of directedness: "In presentation something is presented" means that an intentional being sees its world as made up of items of significance, of objects. Looking at Sally's trampled garden, Fred sees broken kale and lettuce stalks, Sally sees work ahead of her, and the rabbit hopping by sees food. An intentional agent's actions are directed accordingly: Fred promises to repair the damage; Sally feels dejected; the rabbit feasts. As with the representationalists, proponents of this approach see perception as a basic intentional capacity, but not a representational one: to perceive X is to be oriented toward X, to be inclined to notice Xs. When Sally says the door is "purple like red wine" and Fred says that it is more "mauve, like my favourite shirt," they aren't arguing about their private, door representations. They are both perceiving the same door but, for each, the door is significant in different ways and therefore their respective responses to it will be different as well: Sally's experiences draw her to make red wine comparisons, while Fred's incline him toward making mauve shirt ones. The Problem of Intentionality, here, is the problem of explaining what this directedness relation consists in, how it forms, and how the behavior of intentional agents is ultimately grounded in it.

These two different ways of picking up Brentano's insight about the peculiarity of intentionality have informed two different philosophical approaches to thinking about the problem. Along the analytic tradition[2] that has both influenced and been influenced by the current computational-representational paradigm of cognition in the cognitive sciences, the Problem of Intentionality is the problem of explaining how internal representations hook onto the world and how they are distinguished internally from one another. The question of how explicitly representational acts (such as deliberating between whether to pick answer A or B on a multiple-choice exam) guide behavior is taken to be continuous with the way that implicitly representational processes such as perception guide behavior. For this reason, even the low-level activity of sensory neurons is taken as representing the external objects that trigger it. This sensory activity yields increasingly complex perceptual representations—of object wholes and situations—that guide behavior. In what we might call sophisticated intentional systems, such as human beings, these representations are accessible for manipulation in the special ways I've been describing. A human can become aware of his own hunger and even of his own draw toward the lunchroom but decide not to eat—Fred restrains his own pull toward the pizza by reminding himself of his goal to lose weight: "only human animals possess the representational processes necessary for systematically reinterpreting first-order perceptual relations in terms of higher-order, role-governed relational structures akin to those found in a physical symbol system" (Penn, Holyoak, and Povinelli 2008: 111).

Along the phenomenological tradition[3] that has inspired the 5E[4] paradigm in cognitive science, the Problem of Intentionality is the problem of explaining how behavior is directed, how the world becomes present for an agent. This approach is guided by the observation that thought and action are not distinct activities and that analyzing them as such, as the representational approach does, irreparably distorts our understanding of them. Martin Heidegger (1927), instrumental in bringing about this new view, argues that experience is pregnant with *how* the world is present for a subject, and there is no way to step outside of this:

> Taken strictly, this talk about "putting ourselves into such a position" [Sichversetzen] is misleading; for the kind of Being which belongs to such concernful dealings is not one into which we need to put ourselves first. This is the way in which everyday Dasein always *is*: when I open the door, for instance, I use the latch. (Heidegger 1927: 67, 96)

To use a well-worn example, a hammer presents itself to a subject as a thing to be picked up and used in a particular way—it is "ready-to-hand"—and there is nothing beyond this. To be an intentional system is to be always enmeshed in a world of ready-to-hand[5] objects, "equipment." Even when we are engaged in deliberations, when we are actively using our conscious mental capacities to think through possibilities, we are thinking through configurations of an already meaningful world.

On this tradition, then, perception is not an internal representational process at all; rather, it is a subject's ongoing relationship with a world that is continually unfolding as actionable in different ways. We cannot answer the question of how we are able to manipulate representations and use them to create new situations until we answer the more fundamental question of what it means to be a subject for whom the world is present.

Maurice Merleau-Ponty's work (2012/1945) investigates this question by considering the ways in which details of embodiment influence behavior. With respect to the question of intentionality, the important consequence of this shift in focus is the emergence of the idea that all cognition, including perception, is a kind of action. Intentionality, on this view, is the capacity to act in a directed way. Enactivism[6] fleshes these ideas out by providing a non-representational—sensorimotor—account of perception that explains how acting in a directed way is achieved.

Thus, while the analytic tradition analyzes the differences between Fred and Sally's respective perceptions of the purple door as a difference in each person's representation of the door, an internal process, the phenomenologist treats it as a difference in each person's response to the door, an action in the world.

Both traditions are grounded in insights that we must keep as we seek to unravel some of the confusions around intentionality. The analytic insight is that a capacity to represent makes high-level cognitive tasks such as planning and logical reasoning possible. The phenomenological insight is that it is the immediate significance of our environments, not our internal representations of it, that grounds our activity. In other words, we cannot give an account of representational intentionality without first explaining directed intentionality. Precisely how the former depends on the latter will be discussed in more detail in Part II, but to give a sense of how I see these two kinds of intentionality working together, I'll conclude this discussion with another scenario.

Sally is out with some friends listening to a band at the local pub. Maria asks whether anyone wants to order food, and this prompts a discussion over who wants to eat and what food on offer is preferable. Sally, who has recently decided not to eat meat, suggests the roasted mushrooms and tapenade. Maria and a few others decide on the chicken wings and fries, and Fred declares that the deep-fried pickles are the best in town. As they look at the menu and discuss their options, the friends are interacting with their occurrent physical environment—the menu is a physical thing they must pick up and pass around the table—but, as they talk and think about what they might eat, they are also engaging with something not physically present: the culinary possibilities of their immediate future. Eventually, food is ordered and delivered to the table. As they begin to eat, Sally becomes aware of a desire to eat a chicken wing; they smell good to her and she can imagine the tangy, spicy bite of the sauce, the satisfying texture of the meat in her mouth. Maria exclaims, "these are delicious!" and the others agree. Sally feels her vegetarian resolve slipping. To Soraya, a longtime vegan, the smell is revolting, and she is dismayed that it is on the table at all. Meanwhile, Sally is struggling. She decides to leave the table. As she excuses herself and heads to the bathroom, she mentally rehearses the good reasons she has for taking a principled stand on her dietary choices: "my physical well-being does not depend on this food; a moment of gustatory pleasure is soon gone; killing a sentient being is morally reprehensible and while some situations might justify some moral wrongs, there is no such situation here." In this way, Sally uses words—linguistic representations—to literally place reasons in front of herself, reasons that expand the situation and help her resist the immediate sensory draw.

If we view Sally and her situation with a yet wider lens, different aspects of it show up. Sally moves through the situation by acting and reacting to the various features of the ongoing flow. That these features include representations—on the menu, being uttered, "in her head"—is less salient from this vantage point than is the way in which Sally directs herself toward them. In suggesting the mushrooms and tapenade, Sally is steering herself and the others away from the possibility of wings. In seeing and smelling them, she is drawn toward them. She responds to this with a counteraction: she stands up and walks away, and with each step the sensible draw diminishes. Later when the temptation is becoming too great, she uses words to expand the situation to include features that attract her more powerfully than do the sensible ones of the current situation. Because she is more directed toward the added objects—her reasons for becoming a vegetarian—they effectively drown out the objects she is trying to avoid.

By shifting our focus away from the role that representations play in the back-and-forth dance between subject and world, we see that the dance itself is a manifestation of a more fundamental intentionality, a basic bodily directedness toward and away from points of significance. Goal-seeking behavior grows out of this fundamental relation with the world: Sally searches for a refuge *because* the wings are a draw. Artifacts that we describe as though they too were goal seekers—heat-seeking missiles, for example—can behave in one way, fly through the air toward hot areas, but these purposes are not their own and they are merely functioning as they have been designed. When intentional agents act, on the other hand, their purposes *are* their own: Sally searches for a refuge because the wings are a draw *for her*; she wants to resist them.

What does it mean to have a purpose, in this fundamental, non-propositional sense, and how do these purposes motivate/cause intentional agents to act? These questions belong to the Problem of Directed Intentionality. But then we might ask: What makes it possible for some directed-intentional systems such as Sally to use representations to further their goals? How do words function as representations for a directed agent such as Sally? These questions belong to the Problem of Representational Intentionality. In other words, Sally's capacity to see her world as made of objects that mean something to her—directed intentionality—is different from her ability to treat some of the objects in her world as symbols—representational intentionality.

These two questions are rarely separated out in this way; indeed, the practice of treating internal states as representations confuses them and yields a host of intractable problems (grounding and explanatory gap problems, to name just two) that make the Hard Problem even harder. Keeping them distinct, however, brings into focus the relationship between these two capacities: being a representation-user requires that one already sees the world in terms of objects. It requires other factors too: the social and physical environments within which individuals develop play substantive, capacity-determining roles. My task in this book is to give an account of how representational intentionality develops when the requisite internal and external factors are present. But before I can get started, I need to clarify some of the key concepts of the larger theoretical framework within which Externalism is being developed.

2

The Dynamic Mechanistic Framework

As theoretical paradigms tend to change slowly, it helps to situate frameworks within the historical context out of which they emerge. Viewed in this light, the methodological shift we have seen in the cognitive sciences over the past sixty or so years is illuminating.

In the 1960s, a preoccupation with the ontological status of mental properties began to give way to interest in the formal relations between them.[1] Initially, symbolic rule-based models—GOFAI—were used to investigate these relations, an approach that was unsuccessfully challenged by the network-inspired mathematical models of Connectionism (during its second wave) in the 1990s. Despite their lack of success in shifting the general research paradigm, these were intriguing because of their surprising successes: generalizations emerge "for free" out of network structures, and partially degraded networks can still perform well. Most importantly, they shortened the theory-implementation gap, something that was becoming a growing concern for all researchers but especially for those investigating how embodiment details impact cognitive capacities (Brooks 1991A, 1991B). Today, subsymbolic mathematical models known as Deep Neural Networks (DNNs) (Hinton 2007) are the industry standard in AI, and many 5E insights have become mainstream cognitive science.[2] Progress certainly has been made.

The excitement over the recent breakthroughs and successes of DNNs (Parloff 2016) has fueled a surge in computational models of cognitive processes and a swing away, again, from embodiment considerations. As abstract, functional models, however, the usefulness of DNNs in cognitive theory building is limited: implementation details *constrain* functional relations. An information transfer relation, for example, implemented in the wetware of the nervous system, will be subject to and constrained by electrochemical factors and general thermodynamic pressures of decay, while one implemented in a mathematical network will be subject to none of these (Piccinini and Craver 2011: 306). If the local conditions are inhibitory enough, no information transfer will occur. Is inhibition of a particular relation a local *problem* for some information transfer relation? Or is it the proper functioning of a larger mechanism? Sally perceives the meat dishes listed on the menu, but her eyes gloss over them as she brings her attention to the vegetarian options. Her habitual inhibitory impulse to limit information about meat options is a skillful one because it assists her in maintaining a set of challenging dietary restrictions. Thus, while Sally's perceptual system is functioning normally and Sally's perceptual vantage point is optimal, the information transfer is nevertheless inhibited. Viewed from this system-wide vantage point, the wetware constraints that lead to this inhibition are not noise—interferences in an information transfer relation—but part of an information-curation mechanism. If we try to model the perceptual relations in isolation from this larger context, we miss these critical constraints.[3]

More generally, by modeling cognitive functions in isolation from the physical mechanisms that implement them, we put the cart before the horse: which functions are relevant to a given task depends in part upon the structural features of the system enacting them. Functional schemas that model the relations between properties are thus best viewed as "elliptical" mechanistic ones, "as *sketches of mechanisms*, in which some structural aspects of a mechanistic explanation are omitted" (Piccinini and Craver 2011: 284). Given that cognition is a complex bundle of capacities, these implementation details will be many and varied, making increasingly specified, mechanistic models preferable to functional ones.

This is not to say that functional analyses have *no* role to play in theory building. As Levy and Bechtel (2013) convincingly argue, highly abstract models of functional relations play an important theoretical role, even in the cognitive sciences, because they alert us to the *presence* of mechanisms. They cite Uri Alon's (2007) use of graph theory motifs to uncover regularities in cellular processes as a prime example of this synergy between functional and mechanistic analysis. By first identifying the formal structure of *types* of mechanisms, for example, persistence detectors, Alon had a pattern to look for. Only then was he able to uncover the persistence detector processes in the arabinose regulation system of bacteria, an otherwise highly complex bundle of cyclic processes that are very difficult to make sense of. So long as we recognize that identifying patterns is only a part of a larger theoretical task, then, functional analyses play an important role in complex areas such as biology. Indeed, functional and mechanistic explanations should be seen as complementary because they are "inextricably linked: structural descriptions constrain the space of plausible functional descriptions, and functional descriptions are elliptical mechanistic descriptions" (Piccinini and Craver 2011: 307).

The central implication of the requirement that cognitive theories must employ mechanistic explanations is that structural factors cannot be ignored. For a simple example, consider a slingshot, a mechanism for shooting small objects in a straight path with force. Its components are a sling, which can be made of wood, metal, plastic, or some other rigid material, an elastic, and small round pellets. Structural constraints that will influence whether, how, and to what degree a pellet will be propelled by the mechanism include the rigidity of the sling material, the relative weight and shape of the pellets, the strength and resilience of the elastic, the position of the elastic on the sling, the degree of tautness of the elastic as it is pulled back, the angle of the elastic relative to the sling, the release point of the pellet relative to the contraction of the elastic, and so on. By careful consideration of each of these structural and relational details, we can develop a schema of the slingshot mechanism for optimal performance under different conditions. As our paradigm examples of cognitive systems are biological organisms, our cognitive mechanistic explanations will need to be sensitive to the ways in which biological factors influence and constrain our cognitive behavior: "Given that psychological systems are in fact implemented in biological systems, and that such systems are more or less precisely replicated through reproduction, evolution, and development, there are frozen structural constraints on the mechanisms that do, as a matter of fact, implement behavioral and cognitive functions" (Piccinini and Craver 2011: 306).

But physiological structural constraints are not the only ones that impact the cognitive capacities of biological systems, particularly social ones such as human beings: physical and social environmental structures will need to be considered as well. Sensitivity to embodied, enactive, extended, embedded, and ecological factors is the hallmark of 5E cognitive science, and it is within this larger methodological framework that this work is situated. As soon as we acknowledge this need for a wider investigative scope, the complexity of the relations to be considered increases and a tension with the mechanistic methodology emerges. While componential analysis is useful for simple systems such as slingshots in which system output—the propulsion of pellets—is a linear product of the mechanism's activity, it breaks down when identification of individual components is difficult—when system constituents act in highly correlated ways, for example—and when the change they produce is nonlinear—when they produce phenomena that are not functions of system behavior alone, for example. As Kelso and Engstrom (2006) argue, "in the complex systems of coordination dynamics, there are no purely context-independent parts from which to derive a context-independent coordination whole, even though we often try and occasionally succeed to analyze them as such. Again, in coordination dynamics, as in the brain itself, the parts are not one nor are they separate" (202). In such cases, it is more useful to identify patterns, relations between these patterns, and external control factors. We could not give a componential, mechanistic account of tornado formation, for example, because there is no set of identifiable tornado components that always function together to produce tornadoes; rather, there are external factors, wind velocity for example, that *control* tornado formation/dissipation: when the velocity is above a certain threshold, the tornado's funnel formation is maintained; when it slips below, the formation dissipates. Such nonlinear patterns of change defy mechanistic analysis into components because while the control parameter is exerting influence, the individuals caught up in the nonlinear process are "enslaved" and thus do not exhibit their usual behavior. Since it isn't the action of these components per se that is responsible for such phenomena, identification of them is irrelevant to analyses of them. When phenomena are the result of large-scale dynamics, then, mechanistic explanations are neither useful nor practicable.

As our cognitive processes are also biological processes that are part of a system in which these nonlinear factors are at play, no matter how comprehensive the mechanistic account of cognitive activity is, if we don't understand the dynamical relations that impact the behavior of a given cognitive agent, we will "have no explanation of why the organism behaved as it did" (Bechtel 2013: 362). When we recognize that the cognitive explanatory challenge is deepened by the fact that, in addition to the ongoing dynamical effects of the physical environment, social/informational factors both far and near influence, constrain, and subvert local mechanistic processes, we begin to appreciate how difficult it is to even *identify* endogenous cognitive processes as such at all. Indeed, it is possible that some of the factors critical to our cognitive capacities—the ability to identify salient features in a situation, for example—are consequences of autonomic processes that maintain our body's structural integrity. If so, then we will need to rethink the way we carve up the cognitive domain in the first place. Is our directed intentionality a product of a cognitive mechanism? Or is it a product of complex, organizational forces that enable, direct, or constrain cognitive mechanisms?

Bechtel and Abrahamsen (2010) argue convincingly that a "dynamic mechanistic" approach to cognitive theory building is needed to address this challenge. In the next chapter, I summarize some of the key dynamic mechanistic concepts[4] that I draw on in the development of this externalist account of representational intentionality, but I end this section with their own summation:

> Understanding the effects of such organization requires going beyond the approach of basic mechanistic explanation in which scientists attempt to simulate the sequential operation of parts in their heads. Rather, they must invoke a different set of tools, characterizing the mechanism mathematically and performing and analyzing computational simulations. Such an approach integrates the tools of traditional mechanistic science that facilitate decomposing the mechanism and generating basic mechanistic explanations with tools of dynamical analysis that facilitate recomposing the complexly organized mechanism in simulation. This integration of approaches is what Abrahamsen and I refer to as *dynamic mechanistic explanation.* (2010: 366)

3

Dynamic Mechanistic Concepts

To support the challenging task of distinguishing between phenomena that result from nonlinear dynamics and those produced by system mechanisms, we need some conceptual machinery. Here we take our cue from Dynamical Systems Theory, especially as applied to biological analyses. There is much good and important work being done in this area,[1] but in this account of representational intentionality, I will draw only on the twin concepts of self-organization and autonomy.

Self-organizing Autonomous Systems

The second law of thermodynamics tells us that any kind of change in a system releases energy. A consequence of this is that, unless there is some energy intake process, as a system changes over time, it will lose energy: structures of chemical bonds will decohere, and molecular activity will slow down. Maintenance of an organization of particles thus requires energy input: self-organizing systems (SOSs) must *work* to maintain their structure. SOSs are thus thermodynamically open (they freely exchange energy with their environment) and in an ongoing state that is far from equilibrium.

Autonomy, roughly put, is the degree to which an SOS works to prevent dissipation. Following Ruiz-Mirazo and Moreno (2004), we can characterize autonomy more rigorously as " the capacity of a system to *manage* the flow of matter and energy through it so that it can, at the same time, regulate, modify, and control: (i) internal self-constructive processes and (ii) processes of exchange with the environment" (240). Many SOSs lack autonomic resources and are consequently short-lived. A tornado, for example, once formed, is a stable structure that maintains its integrity only so long as control parameters such as wind velocity remain within a certain range. Once these shift, the structure collapses, and the constituents—molecules of water, clumps of sand and dirt, leaves and tree branches, rooftops, and playgrounds—are released from structural control.

To distinguish the special class of autonomous SOS to which living organisms belong, Francisco Varela pioneered the systems concept of autopoiesis: "Autopoietic machines are autonomous; that is, they subordinate all changes to the maintenance of their own organization, independently of how profoundly they may be otherwise transformed in the process" (Varela 1979: 15). In contrast with allopoietic SOSs, whose formation and dissolution are determined by external control forces, autopoietic ones control their organizational integrity from within.

In the context of a discussion on intentionality, autopoiesis is an important bridging concept since it connects our ability to act for reasons with a more fundamental SOS capacity to initiate one's own behavior, namely, to *act* at all.[2] Evan Thompson (2007), developing the work of Maturana and Varela (1980), argues for Autopoietic Enactivism, the thesis that "intentionality arises from the operational closure and interactive dynamics of autopoiesis" (Thompson 2007: 159). Putting aside for a moment the question of how behavior can be *directed*, the idea that the drive to maintain systemic integrity yields a very basic kind of intentional behavior, what I will henceforth call "action-intentional" behavior, is an important one because it serves as a ground of significance for such a system: whatever enables an action-intentional system to maintain system integrity is significant for it; whatever is irrelevant to its autopoietic enaction is not. When the activity of a system does not initiate out of this autopoietic drive, for example, a pedestrian is propelled through the air on being hit by a car, the behavior is not intentional. Systems that have no action mechanism, no autonomic resources at all, are not capable of any intentional behavior and are therefore non-intentional systems. This includes systems having processes that can continue unless perturbed, for example, a pendulum swinging in frictionless space. Examples of non-intentional systems include thermostats, rumbas, DNNs, and automobiles. Strictly speaking, then, only intentional systems *act* since only they can initiate their own activity.

There is more, of course, to be said about intentionality and the role that significance and representation play in that context—this is the topic of the next chapter—but for now I will simply note that, minimally, any capacity for directed or representational intentionality must be grounded in action intentionality.[3] In other words, autopoiesis is a necessary condition of both directed and representational intentionality, although it is not a sufficient one. With these ideas in hand, we can now draw the following distinctions.

Scalar Levels

Systems can be distinguished according to levels of scale along various dimensions with *size* and *time* the most common in theoretical reductions. For biological analyses, it is useful to distinguish *nano* subcellular elements/processes from *micro* cellular elements/processes from *macro* systems and their attributes. Size scale differences typically imply time scale differences since the rates of change of very small particles relative to larger ones is one of fast to slow. At the macrolevel of bodies, for example, not much change occurs in just one second; however, in the same time interval at the microlevel, a single neuron can become active 200–300 times. One consequence of this size/time relationship is that coordinated activity at one size scalar can constitute large shifts in activity at higher levels. Capturing these scalar relations is important in dynamic systems analyses, but for our purposes, distinguishing between micro- and macrolevels of natural systems will be sufficient.

Constitutional and Compositional Levels

Systems are constituted by elements organized at different size scalars, for example, an organism is constituted by cells at the microlevel and organs and bodies at the macrolevel. Constitutional hierarchies thus parallel size/time scalar ones: lower-level constituents are smaller/faster than higher-level ones. The relation between a system and its constituents is constitutive.

Systems, as well as their attributes and processes, can be analyzed along the complementary dimension of composition as well. Indeed, when we aim to give a mechanistic account of some phenomenon, the explanandum (E), we look for the components that work together to produce E. Component hierarchies, however, are orthogonal to constitutional ones; two components that sit at different mechanistic levels may or may not share the same constitutional level. When they do sit at the same constitutional level, the relation between them is causal, for example, one neuron's activation causes the activation of another neuron to which it is synapsed. When components sit at different constitutional levels, the relation between them is constitutive, for example, the thalamic reticular nucleus, which inhibits thalamus activity, is constituted by interconnected neurons. Sometimes these relations are confused, and components are taken to be causally related to one another even when they sit at different constitutional levels, for example, the popular idea that our brains cause us to act in certain ways makes the mistake of treating the brain as causally related to the system of which it is a constituent part.[4]

Following these distinctions, we partition the class of systems into two categories:

Unilevel: System processes are caused by the mechanisms that produce them;
Multilevel: System processes are constituted, wholly or partially, by the mechanisms that produce them.[5]

A DNN is an example of a unilevel system: when a digital image is converted into a vector that initiates a series of intralevel chains of DNN activity, the output produced is fed as input, causally, to other modules—for deciding, planning, or acting. A human being is an example of a multilevel system: when light from an object hits the visual sensors in a human eye, the activated retinal neurons trigger a chain of neural activity that constitutes the system's visual experience.

The affixes "sub" and "sys" will be used to denote micro and macro aspects of multilevel system processes, respectively. Without affixes, the process as a whole is being considered. This aspect terminology should not be taken to entail any ontological distinctions: a multilevel system is one entity, and its constitutional levels are not separate from this whole. An animal's movement from point A to point B involves microlevel neural activity, midlevel muscular contractions, and macrolevel limb extensions. When we want to talk only about the neural processes involved, we can talk about $\text{locomotion}_{\text{sub}}$ processes, and when we want to pick out only limb movement, we can use the locution $\text{locomotion}^{\text{sys}}$ processes. We might also want to indicate the degree to which a process extends across the constitutional levels of a system. When an animal is sleeping, for example, there may be $\text{locomotion}_{\text{sub}}$ activity but no corresponding locomotion^{sys} activity because the motor connections are inhibited; in either case, it is the system as a whole that is moving or not.

Internal, External, and System Attributes

Finally, it will be helpful to clarify the difference between internal, external, and system attributes, as much confusion can arise from a mishandling of these.

Systems have attributes: they are red, blue, hot, cold, and so on. These attributes apply to system wholes. Since a system whole is neither internal nor external to itself, neither are its attributes internal nor external to it. System attributes, however, are things to be explained: they are explananda (Es). When we give mechanistic analyses of them, we cite their components, and it is these that are either internal or external to the system.

When explanations involve a scalar reduction, the components of the mechanism that produce E will be internal to the system. S's temperature, for example, is constituted by the mean kinetic activity of S's molecules. When explanations do not involve a scalar reduction, as when E is the causal consequence of a series of intralevel relations, components might be either internal or external to the system. The attribute of *being powered* as applied to combustion engines, for example, is an effect of the causal relation between a constituent component (internal) and a causal component (external): the fuel is ignited by the activity of the spark plug.

Our capacity for representational intentionality is a system attribute, the result of a mechanism the components of which include both internal constituent elements and external causal elements: the mechanism of directed intentionality in concert with external tools of representation—words, pictures, music—yields our representational intentionality. The name of this account, "Externalism," brings attention to the key role of these external components that are overlooked in reductive accounts. With these distinctions in hand, we are now ready to look more specifically at the representation relation itself.

4

Representation and Significance

Dyadic and Triadic Representation

In the context of cognitive theory building, the concept of representation is a fraught one: much ink has been spilled over whether and to what degree we ought to think about cognition as a representational activity at all (Chemero 2009; Clark and Toribio 1994; Dreyfus 1976, 2007; Freeman 2000; Keijzer 1998; Van Gelder 1995; Wheeler 2004). I have my own position on this topic, which will become very clear as the externalist position unfolds, but here my aim is simply to delineate two very different senses of representation to ensure that no confusion arises because of equivocation between them.[1] It will help to ground this discussion in examples that invoke the two senses I will be distinguishing.

The standard information processing account of perception,[2] on which a system gains information about the macrolevel objects in its sensory environment via neural information processing activity, provides us with a widely used example of one important usage. There are multiple iterations of transductions on this account, processes during which the information of one encoding scheme is converted into another encoding scheme, appropriate to the level of organization handling it. Photons, for example, are transduced into electrochemical signals that cascade across learned/evolved synaptic connections, in this way processing the information they "carry." Bundles of such neural signals are taken as representing the external features correlated with their activation: some visual sensory neurons represent edges, for example, while others represent curves. As the processing continues, different streams of neural activity, each representing different sensory features of the object perceived, together stimulate higher-layered clusters of neural activity, groupings that represent increasingly larger portions of the object, until the entire sensory object is represented. Eventually—and this part is the Hard Problem—the neural macro object representation is transduced into something the organism whole experiences, a mental representation of the object being perceived.[3]

The concept of representation plays an ubiquitous role in the wider biological context as well, to explain relations between mechanisms. Individual cells are self-organized unities that are bounded by a semipermeable membrane that functions to allow the free exchange of some substances between the cell and the surrounding medium and to keep others out. The development of a cellular membrane has different effects. One is unifying. Another is individuating: interaction between cell components and external substances is blocked. When cells are themselves units in a larger self-organized collective such as an organism, however, there are occasions when they must interact with their environments in special ways. Human cellular membranes, for example, do not typically allow compounds such as the hormone epinephrine to cross through their plasma barriers; if such a compound did penetrate, it would be treated as a foreign substance and degraded. The fact that epinephrine is present in the cellular environment, however, is important for the system, the human being, because it is an indication that some activity is being engaged in that requires an increase in some other activity, blood flow, for example. This "organism view" has led systems biologists to invoke information processing concepts such as transduction to explain how and why certain substances are interacted with. On this view, the binding of epinephrine to a cell's receptors "represents" a request for increased blood flow. The response is implemented by activation of the G protein, which in turn activates adenylyl cyclase, which produces cyclic AMP, which is the system that controls the rate of blood flow.

These two uses of the representation concept, which are different from one another in ways I will discuss in Chapter 6, abound in the cognitive science and systems biology literature. They are both importantly different from a third usage we will now consider.

Recall the exchange between Sally and Fred in our opening scenario. Fred tramples on Sally's vegetable bed and, apologizing, Fred promises to help repair what was damaged: "I am so sorry. I will pay for whatever damage I have caused, and I'll help you replant the bed too." Representations play a key role in this unfolding situation. First, Fred uses representations—words—to *mediate* his interaction with Sally: he expands their shared situation with an apology and a promise to repay, directing her attention to them and away from her anger. Sally cannot directly experience Fred's regret herself, but because she is a language user too, she can respond to the representation of it that Fred has put before them. For both Fred and Sally, the initial situation has grown considerably. Where there was only anger, there is now also regret, an apology, and a promise to help and repay.[4]

Fred and Sally seem to be trading in information here in a way that is importantly different from our first two cases: while cellular transductions are direct causal consequences of the physical properties of the vehicles that implement them, Fred and Sally are triggered to respond by *what is being represented*, rather than by the physical properties of the representation vehicles—the words—themselves. Indeed, Fred could have used different words to represent the same sentiment, and he could have used a different mode of representation entirely, for example, a written note or a bouquet of flowers. Whichever physical vehicle of representation Fred uses, Sally's responses are the same. In contrast, the information that blood flow must be increased can be conveyed only by the specific cellular binding process described above.

One important way in which these two senses of representation differ, then, is this: in the cellular case, what is signified is tightly tied to the physical features of the activity that represent it; in the linguistic case, what is signified is not tied at all to the physical features of the symbols that represent it. In the latter, representational success is achieved by something I will for the moment describe as Sally's *understanding* of Fred's words. If Fred were speaking with Sally's dog rather than with Sally, for example, very little to no representational activity would occur.

In his own analyses of representation (Peirce, 1893-1913/1958), C. S. Peirce described this third prong of a representation relation as the "interpretant," the translation of the symbol into the object being represented. Representation, in this sense, is constituted by a triadic relation. Fred's words (X) represent his regret (Y) for Sally (S): S interprets X as an indication of Y. Representation in the cellular sense, however, is constituted by a dyadic relation only. Neural process (X) represents object (Y); epinephrine binding (X) represents the need for increased blood flow (Y): X represents Y. No interpretant is required here because the physical features of X and Y themselves ground the relation. In the triadic case, on the other hand, there is typically no such connection between X and Y; indeed, when symbols are very abstract, there is no physical connection whatever between them. Connecting the two is the role of the interpretant prong. For X to represent Y in this triadic sense, there must be something that "interprets"—in a sense to be explained—X *as* Y.[5]

I will distinguish between these two different senses of representation by specifying the number of the relata implicated by usage: triadic (t-rep) when an interpretant is required for the relation to function representationally and dyadic (d-rep) when it is not.[6]

T-Rep: S interprets X as Y; X represents Y for S
D-Rep: X correlates with/is isomorphic with Y

The reductive hope of the internalist approach to specifying the interpretant prong of the t-rep relation is to show that S's interpretant response is itself a d-rep process. In the next section, I show why this reduction cannot work and then present the externalist alternative. But before we can get to that more detailed analysis of the internalist view, we need one more distinction.

Significance-In and Significance-For

Earlier I noted that action intentionality, the drive of an autopoietic system to maintain system integrity, grounds significance for such a system: whatever enables this activity is significant for it; whatever is irrelevant is not. This very basic significance relation, we might think, is itself the ground of directed intentionality since, as some argue, any system for which the world is "a place of valence, of attraction and repulsion, approach or escape" (Thompson 2007: 158) must already be directed toward its world. We must be careful here. While this insight is certainly correct, this wording of it creates such a tight coupling between the notions of significance and intentionality that it becomes impossible to describe something that can learn to respond to objects in the world with attraction or repulsion without already implicitly assuming some degree of intentionality. Any discussion about whether AI systems are learning systems, then, will already be fraught on this assumption: those who are aware of the implication will reject the idea out of hand, while those who are not will blithely extend intentionality to all learning systems. If we take non-intentional systems such as machine learners to be genuine learning systems, we need a non-intentional way of characterizing the base units of learning, the items of significance, aka "objects," it learns to distinguish from its background environment. To accommodate this, I will distinguish between two senses of significance:

(1) Significance-for: in a way we have yet to explain, intentional agents experience their world as made up of objects and situations toward/away from which their actions are directed;
(2) Significance-in: non-intentional learning systems develop responses to objects and situations but these responses are not directed toward them.

I will not consider here whether, how, and to what degree all perceived objects must be significant-for an S that is an intentional agent; however, when S is not intentional, objects will never be significant-for it; they will always only be significant-in such Ss.[7]

Now there is a rather large elephant in the room here, and it is time to face it head-on. Since the d-rep relations between sensory neural activity and perceptual stimuli certainly do play a role in a system's capacity to discriminate between objects, one might wonder whether they encode significance relations. Indeed, it is this idea that lurks behind the internalist's reductive move: sensory neural processes d-represent those objects and situations in a system's environment that are significant-for it. That these d-rep relations develop at all *means* that the d-represented objects are significant-for such system. Coming to terms with why this significance encoding idea (which is just reductionist representationalism wrapped up in slightly different language) is both insightful and deeply wrong will, I hope, clarify the difference between significance and representation.

Significance and Representation

First, let's consider the insight that d-rep and significance-for relations are deeply entwined. In the typical rabbit's environment, carrots are significant-for it because they are a readily available source of food, and nourishment is necessary for a rabbit's survival. Further, that carrots are significant-for a given rabbit drives its ongoing discrimination between carrots and non-carrots, a practice that yields an increasingly sophisticated discriminatory capacity as the rabbit accumulates new experiences with new carrot varieties in different kinds of situations. The activity/non-activity of the neural pathways that ground the rabbit's discriminatory behavior thus become reliably correlated with the presence/absence of carrots. This correlation between internal sensory processes and external items of significance can be fruitfully treated as a d-representational[8] relation, we might think, because doing so provides a neat theoretical link between the rabbit's microprocesses and its macrobehavior: the rabbit's microprocesses are tuned to carrots because they are significant-for the rabbit. The account thus provides a tool for explaining how neural activity grounds a system in its world.

This seems like progress, so what's the problem? In a nutshell, the idea that d-rep relations encode significance-for relations is overly general and critically so. To see why, we must be guided by phenomenological and enactivist insights.

An intentional agent is always already acting, since it is always concerned with maintaining or attaining some state.[9] Indeed, multiple competing drives are active at any time, and the depth and complexity of these increases according to the extent to which an agent's needs and concerns extend beyond its spatiotemporal present. This hive of activity does not passively perceive everything that is around it and then attempt to sift through the data to determine what might be useful for this or that task; it moves through its world searching, probing, interacting, taking, ignoring, attending. Neither are the "objects" that it ignores or interacts with static things; they too are moving, bending, turning, and, when they are other intentional agents, searching, probing, taking, being driven by their own ends. In the context of a perceiving intentional agent, then, significance-for relations are better thought of as sensory-motor loops—*do* this *with* that—than as d-reps of the objectivist categories our words pick out. The latter divide the world into things according to the bundle of properties they typically have: <*cat*> = "A small animal with fur, four legs, a tail, and claws" (Cambridge Dictionary, May 2024), but the items of significance-for that draw an intentional agent as it moves through its world simply don't match up with these categories. A rabbit's perceptual world might be made up of <*graze*>, <*avoid*>, <*run*>, <*hide*>, while a cat's might include <*chase*>, <*play*>, <*climb*>.

But even these characterizations in terms of what Michael Wheeler (2005) calls "action-oriented representations" (222) isn't quite right since it covers over the way in which what is significant-for is continually shifting, growing to include a wider situation, and then shrinking again. The rabbit is hungry and now its perceptual world is a tapestry of smells that draw it hither and thither. Suddenly another scent breaks through. A movement. The vibration of pounding feet. All enticing smells have disappeared and now there is only the predator's scent on the wind and the intense attunement to the sounds of approach, to the direction from which the threat is coming. There is readiness for flight. As Freeman and Skarda (1990) observe, the neural patterns of activity in a rabbit's olfactory bulb correlate (more or less depending upon how much we are willing to treat as noise!) with the presence/absence of carrots only in certain circumstances and not at all in others. As this is such a critical point, I will quote them at length.

> First, simply presenting an odorant to the system does not lead to any odor-specific activity patterns being formed. Only in motivated animals, that is, only when the odorant is reinforced leading to a behavioral change, do these stereotypical patterns of neural activity take shape. Second, odor-specific activity patterns are dependent on the behavioral response; when we change the reinforcement contingency of a CS we change the patterned activity. Third, patterned neural activity is context dependent: the introduction of a new reinforced odorant to the animal's repertoire leads to changes in the patterns associated with all previously learned odorant. (377)

Why then think of this neural activity as d-representing <*carrot*> or even <*eat*> at all? "There is nothing intrinsically representational about this dynamic process until the observer intrudes. It is the experimenter who infers what the observed activity patterns represent to or in a subject, in order to explain his results to himself" (1990: 377).

When we take a representational stance to the correlations between sensory microprocesses and perceived macro objects, we treat idiosyncratic sensory-motor loops—the significance-for relations that draw an intentional system through its world—as representations of categories, whether objectivist or action-oriented. The subject of such analyses is a relatively static, self-contained, system that divides its world into objects, attends to those objects that will help it meet its goals, makes decisions about which objects present will work best to meet its ends, and so on. As representation-intentional systems, humans do indeed sometimes behave in this way: I may need to go to Toronto and, desiring to do so quickly, I google the traffic flow at various times of the day and choose a starting time that maximizes the likelihood of no traffic.[10] But it doesn't follow from the fact that we are capable of such behavior that we are not also and at the same time agents that, like the rabbit, are continually being drawn hither and thither by their own changing needs and the pushes and pulls of the environment: Sally is searching for a soundtrack to listen to while working; an ad for tents pops up in the sidebar and suddenly Sally is reminded of the camping trip she has planned and her need for a tent and she clicks on the ad; her phone beeps and glancing at it she sees a text from her friend and suddenly she is reminded of the plans she has made for that evening; as she reads the message, the tent, the music, the work, have disappeared. The enactivist insight is that intentional agents are always actively probing and taking and ignoring according to their own autopoietic demands.

The problem, then, with the idea that sensory neural processes d-represent the macro objects (even action-oriented ones) that are significant-for a system is that it covers over how idiosyncratic each incarnation of these processes is. As I've already said, from a high enough level of abstraction, such representational attributions are explanatorily helpful because they connect up microprocesses with macrobehavior. They are ultimately misleading, however, because they entice us toward an overly simplistic picture of the way in which the objects in an intentional agent's world are significant-for it. Thus, while objectivist categories are indeed significant-for language using humans, in ways we shall explore soon, significance-for cannot be captured by objectivist categories.

One final word on the relation between significance and representation more generally. There is a different reason we are tempted to treat these terms as entwined in the way that I have been arguing here covers over important distinctions: sometimes items that are significant-for an intentional system are nomically correlated with other features of the environment, and this correlation makes the one a natural representation of the other. Smoke is a natural sign of fire because fire is always accompanied by smoke. If fire is significant-for S, then S will likely learn to respond to smoke as it does to fire, and we might be inclined to describe this relationship as a representational one: smoke represents fire for S.

As we shall see in more detail shortly, this is too quick. How an object *qua* item of significance is significant-for an intentional agent S is different from how an object *qua* representation is significant-for S: the former rides on an object's physical properties and an agent's needs while the latter rides on an object's representational properties, namely the relational properties that develop within the wider context in which it is used. Sally learns to pick out bananas because they provide nourishment; she learns to pick out "banana" because her linguistic community uses the string to talk about bananas.

One final, final word. The impulse to look within for the representations that we use stems, I think, from our own phenomenological experiences of thinking. Because we can become aware of our own inner states, we assume that it is this reflective capacity itself that turns our inner states into representations *for* us. But, as I argue in the next section, this cannot be right: a t-rep relation is a three-part relation between two objects and an interpreter. A capacity for becoming aware of our own states is certainly a prerequisite for any sort of mental representation-use, but it is not a *sufficient* condition for it. Sentience yields subjective phenomena such as episodic flashes of past experiences, wispy images, emotions, and feelings of balance or scatteredness, but the capacity to treat them as representations requires scaffolding. The business of the next section is to explain what scaffolding is needed.

Part II

The View from Externalism

5

Revisiting the Puzzle

Intentionality

In the previous section, I introduced the central Problem of Intentionality and rehearsed the various ways in which solutions to it have been formulated. The standard formulation is in terms of representation—to think and reason is to represent to ourselves what we are thinking and reasoning about. 5E theorists, in contrast, treat thinking and reasoning as special cases of a more basic intentional relation in which agents direct their actions toward a world that already has significance-for them.

Both approaches are fueled by important insights and so, not surprisingly, neither can stand alone as a comprehensive answer to the question. Representational approaches are problematic, not because of their representationalism, but because of their internalism. In Chapter 6, I will explain why 5E accounts, in shifting the focus toward directed action, address this latter problem; however, they fail to account for our facility with representations. We can and do use representations to great cognitive advantage—we speak, we think, we reason—and any comprehensive cognitive science will need to give an account of this as well.

Externalism begins precisely at this impasse. On it, the Problem of Intentionality separates into the following hierarchy of capacities and related questions:

(1) Action Intentionality: How/why do agents initiate their own actions?
(2) Directed Intentionality: How does the world become significant-for action-intentional agents?
(3) Representational Intentionality: How do directed-intentional agents become robust representation-users, namely, representation-intentional

> agents? What are the necessary and sufficient conditions for robust representation-use?

With respect to the first question, the problem of action initiation, autopoiesis[1] has been a fruitful concept. Within the subfield of AI known as Biological/Enactivist AI, headway has been made on the creation of artificial action-intentional systems by using autopoiesis as a design principle.[2]

The second question is, of course, fraught with hard questions around what it means to be a sentient agent, but we don't need to solve the Hard Problem to make conceptual progress. We might think of a very simple action-intentional system whose interactions with its environment are completely mediated by its embodiment as *directed toward* certain features of its environment in a hardwired sense. In the context of a theory of cognition, however, it is useful to distinguish such hardwired systems from those that *select* what they respond to. A cell's semipermeable membrane protects it from the environment but, since a cell needs an ongoing influx of resources for its own survival, this embodiment detail necessitates an interface mechanism which allows select substances through the barrier. We might call such interface mechanisms "perception systems" since they serve the function of directing the system toward and away from various parts of its world.[3]

A cricket's ability to follow a sensory cue—called "phonotaxis"—might be an example of a capacity on the cusp between being a hardwired, sensory-motor one and the consequence of a very simple perception mechanism. A description of some of the anatomical details of the cricket will help to make this clear. The male's song, a simple tone and a rhythm, is produced by rubbing its wings together. Loudness of song is a feature that females need to be tuned to, since a louder song means a heartier male, and consequently, they have evolved a body that tracks loud songs. Here is how. A cricket's ears are on its forelegs. Neurons connected to each of these ears are sensitive to the vibration amplitude of the signal at each ear: when the amplitude reaches a certain level, the neurons fire and this in turn triggers motor neural activity such that the cricket turns in the direction of the side that fires. The ears are joined by an inner tube that is open to the environment in two places, at each spiracle situated on either foreleg. Auditory sensory input thus arrives at each ear from two places: from outside and from the other ear via the connecting tube. Since the distance between the other ear and the input ear is greater than the distance between outside and the input ear, the amplitude of the sound wave that traveled through the tube from the other ear will be smaller than that coming from the outside. Thus, when a sound originates from a cricket's left side, the large amplitude wave will trigger its motor routine to turn to the left; likewise, when the sound originates from the cricket's right side, it will turn to the right. A cricket never has to decide which way to turn, then, when a potential mate is nearby; its tight sensory-motor loop embodiment ensures that it always tracks in the direction the sound is coming from. Furthermore, since its motor response is directly tied to loudness, it will always track the loudest sound source.

Is this an example of a perception system, that is, a mechanism for resolving ambiguity between potential items of significance? Or is it an example of an ingenious embodiment that ensures that ambiguity never arises? I won't speculate here. Investigating the various capacities of animals along these dimensions is the work of future examination; here the important point is that any action-intentional system that also has a mechanism of perception is, even if only to a small degree, a directed-intentional one. Phenomenological analyses of directed intentionality provide a wealth of helpful data, and this account is much indebted to them.[4]

As I explain in Chapter 4, one of the core insights motivating Externalism is that directed intentionality is not a representational capacity, an idea that seems to be in tension with the reductive representational picture that dominates mainstream cognitive science. My first step toward clarifying the areas of conflict has been to distinguish d-rep from t-rep relations and significance-in from significance-for relations. Representational intentionality, I will argue, becomes possible for directed-intentional agents given certain internal and external conditions. After explaining why an internalist approach won't help us with this question, the meat of the view is presented in two Chapters, 7 and 8, which detail the endogenous factors for a representation-use capacity and the exogenous factors that lead to representational intentionality respectively.

6

Internalism

Before we can give a critique of internalism about our representation-use capacity, we need to distinguish between three kinds of internal representation ascription as only two are relevant to the question of representational intentionality and, of those, each pertains in a different way.

Reductive Internalist Representations

The orthodox account of perception in the cognitive sciences is an information-processing, constructive one,[1] part and parcel of a larger computational/representational picture of the brain-mind. On that view, sensory neural activity, seen as a kind of computation, yields increasingly complex representations of an agent's world, patterns of neural activity that nomically co-vary with the presence/absence of macro objects. In sentient agents, this subpersonal processing sometimes culminates in perceptual experiences that control and guide system-wide actions such as self-reports and deliberations, that is, neural perceptual representations give way, in a sense yet to be explained, to *mental* representations that are experienced by the perceiver. Both neural and mental representations, it is thought, play a critical role in supporting a wide range of higher cognitive activities:

> cognitive representations of events (i.e., of any to be-perceived or to-be-generated incident in the distal environment) subserve not only representational functions (e.g., for perception, imagery, memory, reasoning) but action-related functions as well (e.g., for action planning and initiation). (Hommel et al. 2001: 849)

I will call this kind of representation ascription "cognitive." The externalist view is that such cognitive representations, whether neural or phenomenal, cannot function as relata in a t-rep relation.

As we saw earlier, a different kind of representation ascription is used to describe complex biological processes. Recall that to explain cellular cyclic AMP production in the presence of epinephrine, the concept of transduction is invoked: the binding of epinephrine to a cell's receptors "represents" a request for increased blood flow and is transduced into a "language" the cell can "understand," namely, activation of the G protein. This usage is importantly different from cognitive ones. To explain how, I will draw on Nicholas Shea's (2018) presentation of Varitel Semantics,[2] which does an admirable job of establishing a rigorous method for distinguishing between cognitive and non-cognitive representational appeals.[3]

Systems which perform specific actions—what Shea calls "task functions"—in a way that perseveres in the face of obstacles and over time, require representational explanations that make connections between their behavior and distal effects, connections that cannot be made if we cite only the proximal causes of the behavior. To use an example of Shea's (2018: 124), a rat might perform the task function of reaching a certain location L—in pursuit of a food reward perhaps—in a variety of ways and in the face of obstacles: it might take the shortest route, a circuitous route, or an avoidance route, depending on the features of the situation at hand. By treating the relevant neural isomorphisms that are engaged in the course of this activity as representational components of an algorithm the rat "uses" to perform the task—aka cognitive representations—we not only explain how the rat performs its actions, but we also explain why the rat moves through the maze: to reach L. If we didn't, if we explained the rat's behavior without appeal to such cognitive representations and merely cited the causal chain of events that make it up, for example, neural activity X caused neural activity Y which constitutes the rat moving forward, then we do not connect the outcome of reaching L with the occurrent behavior at all. From the proximal perspective, L is just another point along a causal chain.

> An organism interacts with the environment, bringing about distal effects in its environment, doing so by reacting to distal objects and properties in the environment. There are real patterns in the environment and an agent's

> interactions with it that would be invisible if we looked only at intrinsic properties of the agent. (2018: 32)

Shea's claim here, then, is that the rat's goal-seeking behavior is inexplicable unless we see its relevant inner states as "real" physical algorithms for fulfilling those goals. We should analyze isomorphisms in representational terms when the system in which they are found "exploits" them in order to fulfill a task function that it pursues robustly: "content-constituting correlations are those which unmediatedly explain a system's performance of its task functions" (2018: 110).

More generally, such cognitive representation ascriptions are theoretically productive because they help us predict and explain the behavior of the wider system of which they are a part. Predictive processing models of cognition (Clark 2016; Friston 2008), which make heavy use of the idea that brains function by way of models and hypotheses, appeal to such cognitive representations as well. In mirroring salient features of an organism's world, both concrete and abstract, isomorphic neural activity makes it possible for organisms to hook onto their environments in life-optimizing ways: "That distributed inner model is itself the result of self-organizing dynamics operating at multiple temporal scales, and it functions selectively to expose the agent to the patterns of stimulation that it predicts. The generative model thus functions—just as an enactivist might insist—to enable and maintain structural couplings that serve our needs and that keep us viable" (Clark 2016: 293).

In contrast with cognitive representation ascriptions, the representational gloss of epinephrine binding is not theoretically productive because the explanandum, the production of cyclic AMP, is exhaustively accounted for by the proximal, causal processes that trigger it: the presence of epinephrine activates the G protein which in turn activates adenylyl cyclase which produces cyclic AMP. No wider *goal to increase blood flow* remains to be accounted for since that was merely a heuristic metaphor, not a thing to be explained. These kinds of representational glosses, then, are best thought of as heuristics that help us understand complex processes and not as possible grounds of representational intentionality. They will not concern us further here.

It is cognitive representations, then, that are taken by internalists to be the ground of our representational intentionality: "As I have been describing the problem of mental content, realism about vehicles is a core part of what it takes to be a mental representation" (Shea 2018: 15). The question before us is whether these internal representations can play the role of relata in t-rep relations.

Recall the Problem of Representational Intentionality that we are trying to solve here. At least some of our cognitive behavior is guided by explicit chains of reasoning—deliberations—that invoke representations. Sally sees a bear foraging up ahead and thinks "either I backtrack to the trailhead, or I go off the path here to avoid it." Sally is representing to herself possibilities for action and then choosing from among them. If we want to explain why Sally suddenly stops, turns around, and begins walking back to the beginning of the trail, we will need to cite her choice between these possibilities for action, possibilities that do not exist anywhere except as representations invoked by Sally. In other words, we will need an account of how Sally's behavior is a causal consequence of this representational activity—this is the puzzle. On the internalist approach I've been describing, specific neural patterns of activity play the role of representing the various possibilities that Sally is considering, and it is this lower-level activity that provides the causal force of her action: the neural activity that represents Sally's choice won out and she turns around.

But the idea that the reasons that guide deliberative behavior can be found in an agent's internal mechanisms rests on a confused picture of the origination of agency. That Sally's choice to turn back (C) is an efficient cause of her subsequent actions by way of the lower-level mechanisms that encode C's representational content is a sloppy analysis of how parts contribute to system behavior. Parts of systems do not *cause* system wholes to do anything; they are constitutive of system behavior (Craver and Bechtel 2007). My feet moving forward are not the cause of my forward movement; they are part of what it is for me to move. My lungs contracting and expanding do not cause me to breathe; they are constitutive of my breathing. Likewise, the purported neural mechanisms are constituents of system behavior, not causes of it. For Sally's representations to play a *causal* role in her behavior, they must be macro objects for Sally, and Sally's neural states are not candidates for this. Reductive internalist accounts fail to establish ground for the representational intentional relation, then, because neural processes cannot play the role of X in the t-rep relation.

Furthermore, their approach confuses representation-use with directed intentionality and thus fails to distinguish between deliberative and directed activity: Sally's neural processes might d-represent her possibilities for action—this tells the neuroscientists who map Sally's neural activity to her behavior something about what possibilities Sally is ready to engage in—but it tells us nothing about how her invocation of these possibilities led to this outcome. This is the puzzle of representational intentionality: how do intentional agents *use* representations—deliberate—in these cognitively enhancing ways?

Someone might object here that we have ignored the promissory note implicit in these reductive accounts; once we solve the Hard Problem, we will see that neural representations are themselves constituent parts of mental representations and the latter, as macro objects, *are* potential candidates for X in the t-rep relation. To explore this possibility, I turn now to the non-reductive internalism of higher-order theories of representation.

Non-reductive Internalism

As we've just seen, neural d-representations play a role in most cognitive scientific accounts of perception, but whether and how these yield mental representations are still open questions. For analytic "representationalists," however, that perception involves mental representations is a foundational assumption: to perceive an object, even non-conceptually, is to experience a mental representation with content. For these thinkers, representational intentionality—the capacity to think *about* our world—is grounded in this perceptual representational capacity. As I've been arguing, this initial move is deeply wrong. The representations that we use to reason about our world are macro elements of a t-rep relation—S uses X to represent Y—while the sensory neural activity that is nomically correlated with Xs, if we want to view it representationally at all, is an element of a d-rep relation only—neural activity N d-reps X. For mental representations to play a *t*-rep role, then, they must do so in a non-reductive way. As Locatelli and Wilson (2017) perspicuously put it,

> if we take representationalism to offer a theory of the content of conscious perceptual experiences, and not merely the subpersonal mechanisms involved in perceptual processing, . . . it is necessary not only to show that experiences possess contents, or have accuracy conditions, but that these contents are in some way manifest, or cognitively available, to the perceiving subject at a first-personal level. (209)

Higher Order Perception (HOP) and Thought (HOT) theories of consciousness seek to solve this problem. On these views, representation-intentional systems have special, second-order cognitive capacities—perceptual in HOP theories and mentally cognitive in HOT theories—that enable direct interaction with mental representations themselves. On HOP theories (Lycan 1996), for example, an intentional agent can perceive their own perceptual representations. Likewise, on HOT theories (Rosenthal 2005), whenever an intentional agent is "having" a mental state, perceiving some object for example, they are capable of becoming aware of the mental state itself. On both accounts, the result is a second-order mental representation of which a first-order mental representation is a constituent part. This second-order vantage point turns the original mental representation—perception of Y—into a possible X relatum because now the mental representation is a macro object for S. In conjunction with a reductive internalist account, this will fulfill the promissory note: neural activity N d-reps Y, but it also constitutes mental activity X which represents Y for S.

Considerations of parsimony aside, this "in for a penny, in for a pound" strategy of positing a higher-order representational capacity to explain representational intentionality won't work for two reasons. First, there is a phenomenon/noumena problem here. In a t-rep relation, S treats X as a representation of Y. To do this, S must be capable of distinguishing between X and Y; else S will not know which is playing the role of symbol and which is the object signified and, therefore, won't be treating one as the representation of the other. On these higher-order theories of consciousness, S can distinguish X well enough—on HOP, S perceives X and on HOT, S thinks about X. But what about Y? As on all representational accounts of perception, S's perception of Y is an internal representation of Y, namely X in this example. Even when S is standing in a second-order relation to this perception, the relata are X—S's perception of Y—and X—S's perceptual representation of Y. Since experience is always phenomenal, noumena—the actual objects themselves—are beyond it. Second, the view is empirically unsupported. As anyone who has a meditation practice can attest, while a capacity for briefly reflecting on one's own awareness is minimally available to most humans, *much* explicit development of it is required before it can be maintained long enough to be of any use, certainly before it can be of use in activities such as deliberation. Thus, as a natural phenomenon, awareness of our own ephemeral, wispy, mental flashes cannot ground even the X relatum of a t-rep relation.

Internalism, then, whether reductively or non-reductively construed, cannot account for a t-rep capacity: the former fails because the X it posits cannot serve as a relatum in a t-rep relation, and the latter because the Y it posits cannot serve as a relatum in a t-rep relation. HOT views are also phenomenologically inaccurate: the ability to reflect in the way supposed is a learned skill, one that requires the scaffolding of a community within which there is a practice of representation-use. The externalist account is an explanation of what internal and external scaffolding is required to support the development of this capacity.

Conclusion

Explaining how distal causes such as beliefs become efficient causes is a goal of both internalist and externalist theories of content. But because distal causes are, well, distal, they must yield effects by way of other, proximal, factors. On the internalist approach, these are the physical mechanisms that function as internal d-representations of distal conditions, an idea that, as I've been arguing, cannot work. On the externalist approach, the representations of distal conditions are themselves external. The externalist thus owes an account of how external resources take on this role. I turn to this in the next chapter.

7

The Foundations of Representation-Use

Introduction

As should be clear by now, the externalist approach to the question of representational intentionality is to first distinguish a basic directed intentionality from a capacity for using representations. This clarification and narrowing of the scope of our investigation is helpful because it allows for theoretical development in a difficult area without the added burden of the Hard Problem.[1] Once we are clear on the conditions that ground our representation-use ability, we can investigate what external factors facilitate the emergence of representational intentionality. This chapter is concerned with establishing the first part of this task; namely, investigating the necessary conditions of representation-use. In the following chapter, I take on the second part, examining the external features that fuel the development of a robust representation-use capacity, representational intentionality.

This chapter unfolds in three parts: (1) *Learning and Significance* lays out the necessary conditions for a system to learn significance triads, the foundation for a triadic representation (t-rep) relation; (2) *Expectation* gives an account of the "interpretant" prong of the t-rep relation in terms of expectation; and (3) *Direct and Indirect Representation-Use* explains the difference between two ways of implementing a t-rep relation.

Learning and Significance

To keep this account as general as possible, so that it might apply widely to any system (S) capable of basic learning, I will describe it using simple associationist learning concepts. As many might look askance at this behaviorist approach, I feel compelled to note here that while I agree with the general cognitivist sentiment nicely expressed by Michael Tomasello that "modern developmental psychologists and cognitive scientists no longer think of children's learning as isolated association-making and induction, but rather . . . as integrated with other cognitive and social-cognitive skills" (2005: 3), we must be careful about how we characterize those skills. For example, if "intention-reading (theory of mind, broadly conceived)" (2005: 3) is narrowly conceived as a species of internalist representationalism, then we run headlong into the circularity problems I rehearsed in Chapter 6. Consequently, before we look at the ways in which the rich, multifaceted activity of language use manifests in the social animal context, we must become clear on the necessary and sufficient conditions of representation-use activity narrowly construed. For this task, Associationist Learning Theory is an ideal framework. I start by clarifying some terminology.

A stimulus is something in a system's environment that provokes a response, for example, secretion of saliva, increase in heart rate, movement, perception. As a response-provoking thing, a stimulus is an item of significance-for/-in S, though the degree to which it is significant depends on the strength and breadth of S's learned responses to it. Regularities such as kinds, for example, <*owl*>, those for which we have no labels, for example, low-level features that a deep convolutional neural network discriminates, situations, for example, <*skunk odour in the air*>, are all potential stimuli. A stimulus is unconditioned (US) if it provokes an unconditioned response (UR), what we sometimes call an "instinctive" response. A stimulus is conditioned (CS) if it provokes a response (CR) only after a period of associative learning. Herein I will use the terms "stimulus" and "object" interchangeably to indicate an item of significance-for/-in some S.

A simple, now classic example of conditioning will help ground this discussion. Ivan Pavlov (1927) showed that an unconditioned stimulus-response relationship in dogs, between meat (US) and salivation (UR), could be exploited to establish a new stimulus-response pairing, between a previously neutral stimulus, the buzzing of an electric bell (CS), and UR (CR). After repeated experience of the constant conjunction of US/CS, UR is elicited in situations in which CS alone is present. Thus conditioned dogs thereafter respond to the buzzer (CS) as though to the presence of meat (US) by salivating (CR).

In the laboratory, arbitrary associations of this sort serve as a useful control. But in the wild, associations are seldom arbitrary in this way. Often it is the case that one bundle of features—bundled because they belong to some natural kind and so always appear together—is correlated with another bundle of features, perhaps because there is a causal relationship between the two. To use a well-worn example, fire (instinctively US-aversive for many animals) is almost always accompanied by smoke. As a result of experiencing the constant conjunction of the two,[2] an organism capable of associative learning will, over time, develop a conditioned response to smoke.[3] Such an organism, after learning, will respond to smoke as though to fire, even in the absence of fire. Through its association with something that was already significant to the organism, a new item of significance develops: smoke is now an object, a thing to be noticed.

This relation between US, CS, and UR/CR is what I shall be calling a "significance triad." So far, there is no representation here at all, but this triadic relation between items of significance is its foundation. CS is significant to S because US is, that is, if fire weren't response-provoking, smoke wouldn't be either. This establishes *that* CS is significant, a thing to be noticed. But for the relation to be a representational one, CS must be more than merely significant; it must be related to US for/in S in a particular way. This connection, I argue in the next section, is established via expectation, an ability that itself requires some further scaffolding. I turn to an explication of the expectation prong of the t-rep triad now.

Expectation

The sense of "expectation" being developed here follows the usage established by predictive processing (PP) models of the brain[4] under which bundles of neurons "expect" a certain input when they are primed to respond to that input, that is, when their action potentials increase toward the threshold of activation given that input (Clark 2016; Friston 2008). Expectation is thus an attribute that applies to a system *the constituents of which* can be primed in this sense: S expects X when S is primed to respond to Xs.

A word of clarification is needed here about the proper object of an expectation relation. In the PP literature, one often reads of neurons expecting some situation or object. This is a convenient shorthand for indicating that they are primed in a certain way. But we must be careful about how we describe *what* they expect. Neurons trade in electrochemical impulses—these are their objects. The items of significance that are objects for/in S, on the other hand—these might be cats, tables, birthday parties and the like if S is human—are not things at all at the microlevel. While *we* can track the correlational patterns of activity between a cluster of neurons (N) and the presence of some macrolevel object (X), the neural cluster N itself cannot since X and N exist at different scalar levels. When there is a regular correlation between activation of some N and the presence of Xs, then, the most we can say is that N *d*-represents X in S.[5] For the same reason, if N is primed whenever some other bundle of neural activity, N_1, is active, it would be wrong to say that N_1 expects X (in virtue of the priming and N's d-representation of X): X is not a thing at all at the scalar level at which N and N_1 interact.[6] To ensure that we don't lose sight of the object of expectation we are tracking, I will describe such cross-level priming relationships analogously to cross-level correlational ones, as d-expectations. With these clarifications in mind, I turn back to an explication of the expectation relation.

According to PP, neural processing is hierarchically organized into layers of neural activity. With respect to a particular processing task, for example, the perceptual one of visually identifying an object, the initial retinal sensory activity is conceived as bottom-layer processing, d-representing low level, sensory features of the object, and the clusters of neural activity following it as increasingly higher layers of processing that model bundles of these features until eventually the object whole is d-represented. Sensory information that feeds forward through this network trigger d-expectations for certain objects and situations via higher-layer neural pathway priming. Perception of a table set for dinner, for example, might trigger a d-expectation for a meal. These d-expectations, in turn, feed backward to the sensory front guard, priming neural pathways accordingly, "predicting" what sensory information is present or about to follow. Since the d-expectation pathways are already active, only new information that conflicts with these d-expectations/predictions—these are errors—are processed. A strong smell of gasoline when one is expecting delicious food odors, for example, might draw and keep one's attention as the new information is processed and sense is being made of it. As the synaptic pathways unfold, the action potentials of neurons up and downstream fluctuate accordingly as d-expectations/predictions shift according to the details of the current situation, which includes activity originated by the system itself, for example, visual searching for possible sources of the gasoline smell, verbal inquiry about its source, and so on. Zooming into the activity of an individual neuron, we can see that this larger-scale predictive behavior is grounded in learning, developmental or evolutionary. Once sets of synaptic pathways are established, the activity of a single neuron will prime the neurons along the pathways connected to it. As action potentials admit of degrees in nervous systems, different d-expectations with differing degrees of action potential vie for primacy in this ongoing, recurrent cycle of error and d-expectation/prediction. The primed pathways that win out constitute the possible outcomes that S expects.

Following a tradition in cognitive science that began with the development of connectionist models of learning, I distinguish between two ways in which expectations manifest: implicitly, as neural priming consequences of our synaptic pathways, and explicitly, as episodic flashes, trains of thought, feelings, emotions, and behavior.[7] Just as triggering any part of a neural expectation pathway will prime the rest of the pathway, so too experiencing episodic flashes will prime the rest of its expectation pathway. These fleeting expectation relations cannot themselves function as t-representations, however, as they are too ephemeral to support deliberative cognitive activity, such as planning and problem solving, that requires stable representations. Before a mental trace can function as a t-representation of some object, then, more scaffolding is needed. More on this momentarily. To ensure that we do not begin to think of these different aspects of a single expectation response as ontologically distinct, I will replace the implicit/explicit terminology with $\text{expectation}_{\text{sub}}$—to denote the subsystem *aspect* of an expectation response—and $\text{expectation}^{\text{sys}}$—to denote the system *aspect* of an expectation response. Without scripts, the term denotes the system-wide response. This terminology allows us to unambiguously refer to neural-level activity or system-level activity when the need arises but makes it clear that these are constituent components of one overall response. It also extends the predictive processing account of brain activity to multilevel processes: "powerful forms of learning . . . are able flexibly to combine top-down and bottom-up flows of information within the multilayer [*and multilevel*] cascade" (Clark 2016: 26).

With a PP learning mechanism, the full triadic representation relation can be implemented—the system can learn new associations while maintaining expectation associations (priming) for old ones—and representation-use becomes a possibility. When S responds to CS with CR/UR, the pathways for US are primed and there is an expectation for US. When S is also a multilevel system, these expectations cross levels of organization, not as causes, but constitutively they suffuse through the system: during an expectation of danger, specific neural processes become active, pupils dilate, respiration increases, a general alertness to danger takes hold. Phenomenologically, something appropriate only to sentient systems, expectation^{sys} is manifested as the pull to do whatever can be done in a given situation, to use what is "ready-to-hand." Expectations^{sys} "tune" and direct an organism to the salient possibilities in its environment; if a usable tool is present, that tool will be noticed. As Dreyfus, à la Merleau-Ponty, puts it, "as I cope with a specific task in a specific situation, other situations that have in the past been relevant are right now present on the horizon of my experience as potentially (not merely possibly) relevant to my current situation" (2007: 1158).

Let me summarize what has been established thus far. Representation-use is grounded in what I have been calling "significance triads": (potential representation-user) S learns to respond to (representation) CS *as though to* (object represented) US. Pavlov's dog responds to the buzzing bell *as though to* meat. While there is a promising mirroring here of the triadic representation relation, it falls short if S's response to CS is not accompanied by an expectation for US. Without this, the response is not *as though to* US; it's just a new learned stimulus-response pairing and the triadic relation collapses into a dyadic one between CS and CR. When it *is* accompanied by expectation for US, however, the triadic relation is established. If S responds to CS with CR/UR while expecting US, CS t-represents US for/in S. We are close to identifying the necessary conditions for a representation-use capacity, but we are not quite there: the *degree* to which S expects US modulates S's representation-use along a direct/indirect dimension, and this has important consequences for a system's representational intentionality.

Direct/Indirect Representation-Use

Whether the relata of a t-rep triad are macro objects or a mix of micro/macro elements is dictated by the degree of expectation supporting the third prong:

when S's response to CS as though to US includes only an expectation$_{sub}$ for US, the relata are sensory triggers and primed neural pathways and S's t-rep response to CS is best described as one of indirect representation-use; when S's response to CS as though to US includes an expectation^{sys} for US, the relata are potential objects of interaction for S and S's t-rep response is a direct representation-use of CS. In other words, the degree to which S's priming for US crosses its levels of organization modulates whether its use of CS is direct or indirect. Given that a system has limited resources, the extent and stability of its expectation priming will be impacted by other draws on system resources. In a dynamic system such as an organism that is continually responding at various levels to the dual demands of its needs and the ever-changing environment, the degree to which it expects one thing will expand and contract as other expectations compete with system resources. Systems that are not sentient will never be capable of expectation^{sys} of course; nevertheless, even in such systems there will be priming that constitutes system-level behavior changes—drooling—and priming that does not—localized neural activity. To the extent that expectation extends across the system's levels of organization, to that extent the representation-use behavior is direct. Since expectation for a given stimulus is thus an unstable, expanding and contracting system attribute, representation-use behavior too is unstable. For this reason, maintaining a stable expectation^{sys} for US is a necessary requirement of robust representation-use behavior. In the next chapter, we will look at the way in which a cultural practice of representation-use serves this expectation^{sys}-stabilizing end, but now I'll clarify this direct/indirect representation-use continuum with some examples.

Crickets[8] have been shown to be capable of learning simple t-rep relations (Mizunami et al. 2009): the cricket's d-rep water pathways are primed (US is d-expected[9]) when a banana odor (CS) is presented after conditioning between CS and water (US) reward. Such a response to banana odor certainly looks to be a t-rep one—cricket S responds to CS as though to US while expecting US. But because all of a cricket's responses to CS and US are indirect, via d-representation and d-expectation relations, the cricket's t-rep response is indirect as well. Put in phenomenological terms, since CS and US are ready-to-be-inferred objects only for the cricket, learning to expect$_{sub}$ the one in the presence of the other won't thereby make them ready-to-hand. Objects that are ready-to-be-inferred are not things-to-be-grabbed; rather, they are potential targets of action routines—e.g., rotate 40 degrees; extend feelers until contact is made—that are enacted when the relevant neural pathways are activated.

Ants (and bees) provide another interesting case because their colony-driven behavior can look like direct representation-use behavior and it is often described as such in the literature (Jackson and Ratnieks 2006; Hölldobler 1999; Piqueret and d'Ettorre 2021; Schönrogge et al. 2017). Working through an example will help us decide whether such ascriptions are warranted or not.

Imagine that there is a patch of honey (US) in the vicinity of an ant (S). S's olfactory sensors pick it up and, becoming active, trigger a food-investigation motor routine (UR), for example, *<move toward scent>*—*<collect>*—*<return to colony while depositing pheromone trail* (X)>. X, in turn, acts as a trigger of the food-investigation motor routine (UR) for other ants that encounter it. This behavior looks representational: pheromone trail (X) represents the location of the food source (Y) to other ants (S) by way of the UR it elicits. Let's take a closer look.

Fellow colony ant (S_1) moves from the nest and senses X, the pheromone trail left by S. X triggers UR in S_1: it moves toward the patch of honey while being primed/alert for Y. There are likely nuances here: certain pheromone trails might trigger expectation$_{sub}$ for specific scents, which in turn make up specific kinds of food, or a pheromone trail might be more general, triggering expectation for a range of scents which narrow to a particular pathway only once sensory information from the actual source is gained. Whatever the case for a given S, however, the sense in which the macro object Y *<food source here>* is a target for an ant is indirect only: an ant's landscape of objects are all things that are ready-to-be-inferred, not ready-to-hand.

Thus, while expectations$_{sub}$ ground the t-rep relations that ensure that S's behavior is optimally adapted to its environment in these complex ways, they cannot yield opportunities for *direct* representation-use and thus cannot serve as a ground of representational intentionality. For an S that expects^{sys} US when confronted by CS, however, the relata of the t-rep triad are potential ready-to-hand objects for S and direct representation-use becomes possible. A vervet monkey hears a troupe mate utter "rraup," an aerial call that might signify something like *<danger from above>*.[10] Expecting^{sys} the presence of a predator such as an eagle—episodic memory flashes of past encounters with birds of prey, feelings of fear, rapid pulse—it moves immediately to high alert and looks up, scanning the treetops, all the while moving to ground, away from the potential danger. As the monkey moves through its environment gathering pertinent information about the current situation, new expectations arise and, with them, new sensorimotor impulses. The more ready-to-hand objects the environment provides, the more possibilities for response there will be. Hearing the <*danger from above*> call, a vervet monkey might be drawn to emit its own loud "rraup," move/hide, look, make another call, and so on. A chimpanzee, hearing a snake alarm call, will likewise shift to a heightened sensorimotor alertness and scan for the danger but, having more tools at her disposal—a wide range of gestures and vocalizations—a female chimp might also gesture to her children to <*climb-on*> and "hoot" to those who seem unaware of the danger (Schel et al. 2013; Slocombe et al. 2022).

Directed-intentional systems are thus naturally endowed with the capacity to exploit the significatory relations in their environments, not by having internal *representational* resources, but by being deeply "tuned" to their environments in the way I've been describing. The more robust a system's capacity for learning new behavior, the wider the range of possible responses there will be. The consequence is a flexibility continuum of direct representation-use activity, with innate, fixed, alarm call activity at one end—for example, the relatively inflexible mimicry calls and camouflage techniques of mammals, reptiles, and some cephalopods—and learned, variable, language activity at the other—for example, the abstract gestures and vocalizations of chimpanzees and bonobos and the vocal clicking of dolphins (Beecher 2021; Bergman et al. 2019; Culbert and Balshine 2019; Fedurek 2016; Fitch 2020; Grüter and Czaczkes 2019; Sayigh et al. 2023; Searcy 2019).

Seen through this lens, linguistic sophistication is not the genetic inheritance of a specialized language module, but the explosive feedback effect of the ongoing *use* of increasingly abstract representational tools in a social context. The work that phenomenologists and 5E cognitive scientists have been doing to fill in this picture is critical. To set the stage for the externalist account of how it is this confluence of internal and external factors amid an evolving representation-filled landscape that yields representational intentionality, I offer a speculative description of how conventional symbol use *might* have developed. The externalist account does not hinge on this fanciful picture being accurate and it is not being put forth as a hypothesis about how language actually got its foothold in *homo sapiens* culture; rather, it is intended as an intuition pump, something that will prompt us to think more widely about the external conditions necessary for the development of a linguistic cognitive niche once representation-use has become a possibility.

Imagine an early *hominin* (S), sentient to some degree, standing on a ridge atop a broad valley. A movement in the distance catches S's visual attention: a large swath of orange moves along as clouds of gray-black smoke billow above it. Fire. As there is little wind, S's family group in the valley below is as yet completely unaware of the fire. Seeing this, S is moved to warn them. At the same time, S is responding (UR) to the danger itself: adrenaline is pumping through S's body, his pulse is rapid, and his senses are highly tuned to his immediate surroundings. S is preparing for flight. He is also in a broad state of heightened expectation for a wide range of behavior. Experientially, this is manifested as an attentional pulling toward various aspects of his environment, with certain features of S's surroundings coming into sharp relief against others that fade into the background. As S turns and moves, the space of expectations and possibilities for action continually shifts: if a tool is approached or handled, the significance relations associated with it become central, and other relations that had been in expectation diminish. If S drops a tool or is distracted by a sharp sound, yet another set of relations becomes heightened. Through this fluid, dynamic dance, the possibilities of the current situation unfold before S.

The remains of a fire smolder nearby. S has learned that smoke is a natural sign of fire and is drawn to it; it is ready-to-hand for S. S picks up one of the smoldering pieces of wood and waves it overhead. As S moves, waving the burning log, the situation changes, and new sets of expectations become dominant. Thus is S drawn further along and through the situation.

Let us stipulate here that S has no attendant thoughts—he is so "caught up" in the situation, so present, that, even were it possible, there is no conscious thinking, no "I am using this to do this." Nevertheless, S's smoke-waving activity is an example of direct representation-use, albeit a weak/unstable one. S is *using*$_{sub}$, we might say, smoke to evoke the response that fire generally triggers. As he waves the smoldering log above his head to warn others of danger from fire, the smoke is acting as a sign (X) of the danger (Y) itself. Many of the animals that S passes have not learned the significatory relationship between smoke and fire. They respond only to the physical features of S's presence, for example, his rapid movement, perhaps scurrying out of the way as he approaches. When S's family in the valley become aware of him, however, they respond to the smoke (X) as though to fire (Y): it spurs them to action (UR) and, as they move to safe, high ground, they keep wary eyes on the horizon, looking for the fire.

We might wonder whether some conceptual scaffolding is necessary before S can be drawn to behave in this representation-using way. Does the fact that S has learned that smoke signifies fire by itself explain S's stick-waving actions, or do we need to assume in addition that S has a theory of mind, an understanding that his own response to smoke will be mirrored by others like him?

Seen as the result of a conscious decision to act, the move to use smoke in this significatory way does seem to require a premise that links S's actions up with the potential actions of his kin. But, on Externalism, S is not engaged in reasoning here at all. Prelinguistic representation-use results from an impulse to act, to use the tools that are ready-to-hand, not from ratiocination. Reasons and reason*ings* require far more scaffolding than the world that S inhabits offers.

Human children (and adults too) sometimes act in analogously arational ways. When a child's vision is obstructed by something, say a blanket draped over their head, they act as though nothing can see *them*: because they can't see, they can't be seen. As a piece of reasoning, this is poor. For the premise "I cannot see anyone" to lead to the conclusion "No one can see me," there must be an intermediate premise something like "When one cannot see, one is likewise invisible to others" for which, being patently false, the child cannot have evidence. A more parsimonious explanation is that neither the child nor S is engaging in reasoning at all. S is drawn to create smoke because S responds to smoke and thus smoke is ready-to-hand for S; the child acts as though they can't be seen because they cannot see, and thus obstructed vision is ready-to-hand for the child. In both cases, the actions are straightforward responses to what is available here and now.

Let us step back and see where we are now. First, I discussed how the natural conjunction of features in the world could result in learned responses to otherwise neutral stimuli such as smoke. I called these significance triads: S responds to CS with CR/UR, namely, as though to US. Unless the system can retain an expectation for US, however, the response relation will collapse into a simple dyadic one, between stimulus and response: S responds to CS with CR and US/UR drops out. Expectation is the attribute that keeps the relation triadic and, thereby, a t-representational one: S responds to CS with CR/UR *while expecting US.* When only $\text{expectations}_{sub}$ for US are possible, CS can be used only indirectly as a representation of US; when $\text{expectations}^{sys}$, for US are possible, however, direct representation-use of CS becomes possible.

In phenomenological terms, heightened expectations direct what is ready-to-hand in the environment. But direct representation-use is a spectrum capacity, and there is a wide gulf between the simple, relatively inflexible, representation-use activity of animal calls and the sophisticated, flexible, representational activity of language users. Further, as action-intentional systems are continually readjusting their goals in response to a changing landscape, $\text{expectations}^{sys}$ too are continually shifting. Representation-use behavior in this context is fleeting and unstable, sustaining only short bursts of direct representation-use activity—the chimpanzee gestures to her young—not the stable, deliberative activity that is the mark of representational intentionality. I turn now to the development of the language tool itself, the ground of the robust, direct representation-use capacity we are seeking to understand.

8

The Emergence of Representational Intentionality

Introduction

The goal of this chapter is to show how representational intentionality grows out of representation-use behavior given fertile conditions. The account builds on the following four observations, the first three of which have already been discussed.

(1) As action-intentional systems, some of our behavior is aimed at our own ends; we act.

(2) As directed-intentional systems, we act and we perceive: our world is made up of ready-to-hand objects.

(3) Expectations^{sys} for what is ready-to-hand—mental imagery, feelings, system-wide states—turn significance triads into t-rep relations. These expectations^{sys}, however, are themselves too ephemeral and insubstantial to serve as stable representations of intentions—reasons—that could act as causes of our actions; at most they direct us as constituent parts of our actions.

4) Tools of representation—words—serve as stable signs of what is significant-for us. When our actions are intentionally directed toward or away from these, our behavior is reason-guided, namely, representation-intentional.

This idea, that our representational intentionality emerges out of the codevelopment of language and language use, parallels a complementary syntactic view on which our grammatical skill emerges out of the codevelopment of language and language use. Such usage-based, linguistic arguments have gained momentum recently, and as I will be appealing to their plausibility in the second half of this chapter, I will briefly overview here the nature-nurture debate and present some of the arguments that have tipped the scales in favor of a learned capacity. I will then argue that representational intentionality emerges out of this developmental progression from rudimentary to sophisticated language user in the context of a linguistic community.

The Language Tool

As has been observed (Chomsky 1968; Davidson 1975; Hockett 1977), language can be rigorously distinguished from animal signaling of the sort I've been describing thus far by, at least, two key features: compositionality and productivity. Animal communication systems seem to be made up of units that can neither decompose nor be combined, or, if they can, only to limited degrees and in fixed ways. Likewise, they do not seem to be productive, with the repertoire of calls for a given animal being more or less fixed for its lifetime. These observations, some claim, support the internalist hypothesis that some innate factor—a language module—must account for language development. Chomsky-inspired theorists take this position. Hauser, Chomsky, and Fitch (2002), for example, argue that an innate recursive capacity is required to explain our productive use of language.

In direct challenge to these claims, Yosef Prat (2019) argues that there is, in fact, little to no *empirical evidence* that human language is importantly different from that of non-humans: "the vast scientific efforts over many years to track the evolution of this prominent human trait, using comparisons with other animals, have failed to find a definite distinction in this regard between humans and nonhuman animals. If anything, the accumulated empirical data imply the contrary: Many nonhuman animals 'talk' like humans" (885).

Further, humans might not possess a capacity for recursion at all. Plausibly, a recursive capacity entails a capacity for decomposition, the ability to pick out the syntactic units that together compose the whole. If humans are deficient in this, recursive proficiency would thereby be implicated as well. A study (Van Leeuwen, Verstijnen, and Hekkeit 1999) designed to investigate precisely this showed that humans are singularly poor at decomposition tasks: while we are adept at mentally combining images—imagine a horse with a horn—we are poor at mentally decomposing them—rotate the bottom right quadrant of an image you are remembering. The study further showed that to succeed in decomposition tasks, humans require continual, sensorimotor interaction with the object they are considering. This, they argue, is evidence that a capacity for recursion may be a learned skill rather than a native feature of our brains.

Picking up on this provocative suggestion, Christiansen and Chater (2015) argue that not only is an evolved recursion mechanism for language processing highly implausible—the features of a living language change too radically and too rapidly to constitute a stable target for evolution—the Universal Recursion Hypothesis cannot easily explain why recursion isn't a feature of *all* human languages (Evans and Levison 2009), why typically only specific recursive structures (tail recursions) are seen in human languages while others (complex middle recursions) are not and are extremely difficult for humans to process (Christiansen and Chater 2015: 3), and why deficits in recursive abilities are improved with training (Roth 1984; Wells et al. 2009).

To explain these complementary neural empirical findings, Christiansen and Chater (2015) propose that a more general ability to detect sequence patterns, to extract "regularities from highly complex sequentially organized input" (2015: 4)—an ability we share with many animals including nonhuman primates, birds, and rodents—selects for languages that have "recursive consistency" (2015: 6). Using domain-general, simple recurrent networks (SRNs) that have been trained to predict sequences, they demonstrate that such a sequential learning capacity does indeed select for recursively structured languages: "the simulation results provide an existence proof that recursive structure can emerge in natural language by way of cultural evolution in the absence of language-specific constraints" (2015: 6). For systems with a sequencing capacity, it seems that systematic languages are easier to *learn*. Reeder (2004) demonstrates that humans show the same pattern of preference for recursively consistent artificial languages that the SRNs do. Together these results strongly support the usage-based language hypothesis that the syntactic structure of language emerges out of the symbol-use behavior of language users having these domain-general cognitive capacities: "When human beings use symbols to communicate with one another,[1] stringing them together into sequences, patterns of use emerge and become consolidated into grammatical constructions" (Tomasello 2005: 5).

The lesson of these findings takes us to a move away from the syntax-first focus that has dominated computational accounts of cognition over the last five or so decades.[2] 5E insights, such as those highlighted in Tomasello (2005), ground our investigation in use *practices* and, in doing so, make possible the observation that our linguistic propensity is predicated on a more basic social one. With this wider lens, we can look for a more general set of capacities that both support our social nature and lead to language building activities: "the ability to share attention . . . (Bakeman and Adamson 1984); the ability to follow the attention and gesturing of other persons to distal objects and events outside the immediate interaction (Corkum and Moore 1995); the ability to actively direct the attention of others to distal objects by pointing, showing, and using other non-linguistic gestures (Bates 1979)" (Tomasello 2005: 3).

But while internal factors such as these certainly play a crucial role in the development of language, we run the risk of becoming sidelined by reductive internalism again if we focus solely on capacity-based explanations. As Christiansen and Chater (2015) so effectively show, we must look at the subtle and gradual *codevelopment* of signs and sign-use behavior. Consider: the more specific a symbol is, that is, the more it is tied to a particular response, the more constraints there will be for usage and, consequently, the less useful it will be in combination with other signs. I can use smoke, as a sign, in several ways, but those ways are quite constrained toward signaling emergencies. The ease with which a sign vehicle can be produced must also play a role here. Producing smoke is a rather laborious process. If I want to combine this sign with another, also difficult to produce, combining the two would augment those difficulties, making combining less practicable and therefore less likely something I will choose to do. The fewer usage constraints there are, and the easier sign vehicles are to produce, on the other hand, the more a user will be drawn to combine them.

For humans, words are ideal in this regard. They can be produced with very little effort and without the addition of matter. Merely by speaking to oneself or by vibrating the vocal cords, we create word-objects. Indeed, learning to use them is learning to produce them.[3] Imagine if learning to use a hammer entailed creating one! But with each word use, the number of objects in our environment increases and thereby creates more opportunities for responses. This proliferation of word-objects is a key factor in the emergence of representational intentionality, more on which momentarily.

Some excellent work has been done to begin laying down the theoretical framework within which we can better investigate the complex, dynamic processes that govern language development and use. Deacon (2011), for example, has introduced some new concepts to help us theorize about how entirely new kinds of processes emerge out of complex, cyclic dynamics of this sort. Jeffares (2010) argues that our capacity to develop cognitive-enhancing tools rests on the scaffolding provided by earlier cycles of more primitive tool use. From the other side of the fence, Christiansen and Chater (2008) argue that language itself should be seen as an "organism" that evolves in response to selectional pressures.

The field of Evolutionary Linguistics that developed out of these ideas turns the Universal Grammar Hypothesis on its head by investigating which properties of *languages* facilitate language learning rather than seeking to uncover the native linguistic capacities of humans (Kirby, Dowman, and Griffiths 2007; Kirby, Cornish, and Smith 2008). Key discoveries of the field are that highly systematic languages are easier to learn, both for human subjects and artificial neural network agents that have no innate capacities at all (Carr et al. 2017; Galke, Ram, and Raviv 2023; Kirby, Griffiths, and Smith 2014; Raviv, de Heer Kloots, and Meyer 2021), that the systematic structure of language is itself the result of iterated learning over time (Kirby 2002; Kirby, Smith, and Brighton 2004; Kirby, Griffiths, and Smith 2014), and that the larger a linguistic community is, the more structured its languages tend to be (Lupyan and Dale 2010; Raviv, Meyer, and Lev-Ari 2019).

Together, these findings support the externalist hypothesis that, given a representation-use capacity, whether and to what degree language functions as a cognitive support depends in large part upon the external features of the language and the linguistic community itself. In other words, to understand our facility with mental representations and our capacity for reasoning, we need to first investigate how we develop a facility with physical ones.

As in the smoke/fire example I've been using, a natural sign might be some phenomenon that naturally co-occurs with an unconditioned stimulus, or it might be some phenomenon that naturally co-occurs with an unconditioned response. Some animal alarm calls seem to work like this. Flagging, when an animal lifts its tail and displays the white fur beneath, serves to trigger the flight response in nearby animals, presumably because the white fur is also exposed during the flight response itself, as the troupe of animals is fleeing.

Artificial signs, in contrast, can be arbitrarily abstract, ranging from those that closely mirror some aspect of what is represented—for example, *<climb on>* gestures—to those that share no features with what is signified at all—for example, the word "love."

Natural signs such as flagging and alarm calls are often innately driven significatory actions and, as such, indicate an inflexible representation-use capacity. Such systems have some representational tools at hand, but the size of their toolboxes and their contents is fixed. Clearly, the degree to which a system can learn new behavior will impact whether and to what degree it can expand its repertoire of representation tools. Likely, this capacity will also modulate the degree of abstraction it can accommodate. Some factors influence this capacity directly—brain size, neural plasticity, and manipulability of body parts, for example—while others do so indirectly, by influencing the motivation to learn new behaviors. Much could be said here on both scores, but I will leave these details to future fleshing out. Of importance here is the observation that a system might be more or less capable of learning new behavior. Rats can learn to press a lever; a human child can learn to speak; a young chimpanzee can learn to use a stick to "fish" for termites; a crow can learn a novel call; an otter can learn to use a stone as a hammer; a dolphin can learn to spin; and an artificial neural network can learn to respond to an input class with a new output. Provided that representation-use conditions are met, such systems should also be capable of learning to use representation tools to some extent.

The more abstract a sign is, the more flexible this capacity to learn new behavior must be. Bonobos, for example, have a wide repertoire of gestures and vocalizations, some of which seem to be grounded in a natural relation, for example, alarm calling, and others that seem to be more arbitrary, locally-learned, ontogenetic ritualizations. Infant-mother dyads, for example, will develop idiosyncratic, stylized carry signals—shoulder touches—that trigger responses—carrying behavior—that would normally be triggered by overt carry-request behavior—climbing onto mother's back (Halina, Rossano, and Tomasello 2013). Initially, perhaps, gestural and vocal signs are closely linked, sensorily, to their objects; they are naturally co-occurring features of stimulus or response situations. But, as in the case of bonobos, increasingly abstract shorthands for these natural signs develop. The use of onomatopoeic words and stylized gestures, for example, takes us one step further along the abstraction continuum; increasingly arbitrary signs follow.

In the same way that an agent learns to respond in different ways to different objects in its natural environment—through repeated experience with the constant conjunction of the relevant stimulus/response pairs—so too such an agent learns to respond to the different abstract objects, words, in its social environment. A person develops, over time and within a linguistic community within which there is a *practice* of using words in particular ways, a response association between words and the objects to which they refer.[4]

Initially, when infants and small children are still learning to become representation-users, a great deal of both explicit and implicit teaching/learning experience is required to provide the significance framework for perceiving words as objects.[5] We now have good evidence that the more of this linguistic teaching/learning experience a child has, the more sophisticated her language comprehension and use will be; and, the stronger many other cognitive skills will be as well (Haak, Downer, and Reeve 2012; Sénéchal and LeFevre 2002; Suskind 2015; Swanson, Orosco, and Lussier 2015; Yeong, Fletcher, and Bayliss 2017). Mother, pointing to the cat says, "Cat. Look at the *Cat.* Do you see the *Cat*? What colour is the *Cat*?" Since word associations are made possible by the extended linguistic context, individual usage of words will become increasingly constrained to meet the standard that emerges. "Look, there's a dog," says Sally. "That's a cat, not a dog," corrects her mother. The meaning of a word—its "content" in the analytic philosophical locution—is the abstract, spatiotemporally extended pattern of use-response possibilities that emerges out of this linguistic practice.

As a child's linguistic development proceeds, straightforward associations between words and natural kinds—"cat"/<*cat*>—can give way to more complex ones involving abstract objects and situations through combinations of already learned sign tools. The experiences and expectations that underwrite these increasingly abstract learning situations will be dictated by pragmatics—e.g., the motivation for learning, the context of the learning environment—as much as by the associations being learned. Learning to use the word "symmetry" in a logical context, for example, will involve associating it with paradigms of correct logical usage and application situations that will have been experienced by the learner.[6]

The larger a child's toolbox of words is, the greater the possibility for new words will be and the more spatiotemporally distant objects will become potential referents. Seen in this light, signs are a tool for bringing what is neither spatially nor temporally present online. The classic Marshmallow Test (Mischel and Ebbesen 1970) is a stark demonstration of the cognitive advantage this confers on the representation-user. In the typical experimental setup, a child is presented with a marshmallow, an object that exerts a strong sensory pull. The child is told that she is welcome to eat the marshmallow but that, if she can wait for X minutes without eating it, she will receive two marshmallows upon the experimenter's return. Very young children typically give in right away to the sensory allure of the marshmallow: the sweet smell and soft squishability of it are simply too much of a here-and-now draw for the child to resist. Older children, however, fare better. It is instructive to watch them (filmed by a hidden camera) to see what strategies they use to overcome the sensory draw. Self-talk is a critical component of success here.[7] Reminding oneself of future gain—"I will get two in just a few minutes"—makes occurrent an even more palatable situation, thus giving the verbal child a choice where the preverbal child had none.

Recall that one of the central assumptions being challenged here is that phenomenal awareness is itself a mental representation. By bringing attention to the subtle ways in which we lean on what is ready-to-hand in the course of our representation-use activity, Externalism seeks to uncover the scaffolding required for some bit of awareness to become a *thing* at all. For Sally's perceptual experience of an apple to function as a t-representation of <*apple*> or even <*eat-this*>, Sally must anchor her various, idiosyncratic, temporally distinct apple experiences in something fixed. As Wittgenstein (1953/2009: sections 243–279) aptly observes, because experiences are ephemeral and thus cannot themselves serve as comparators, this is not something Sally, or anyone, can do without the aid of an external resource. Words, pictures, and gestures as established objects can serve this role. By associating the same word with a range of ephemeral experiences, the impossible task of *thinking about apples* by conjuring up myriad, fleeting apple experiences (an activity that is bound to lead to mental wandering) is offloaded onto something fixed: "apple." These anchors also function to keep our thoughts focused: if attention has shifted, repeating the word will bring it back. Sally's apple experiences emerge out of lower-level response patterns, to be sure, but they take on a representational role for Sally only in the context of the larger system within which her experiences can be associated with concept anchors, for example, words, in concert with the ongoing activity of the symbol-using community.

This larger system, of course, takes us well beyond the blood-brain barrier of individual agents.

On Externalism, then, mental phenomena are epiphenomenal with respect to a representational function until they are anchored to a fixed object. Lower-level mechanisms implement the representation-use behavior of sentient systems, but they cannot do double duty and account for its representational relations as well. As an individual's linguistic skills become increasingly sophisticated, so too their world expands. It is the combination of this skill in the context of this rich landscape, replete with word-objects at every turn, that fuels a child's developing/emerging representational intentionality.

Representational Intentionality

Consider a typical interaction with Kanzi,[8] a 37-year-old bonobo who has learned to respond to hundreds of words and can use lexigrams[9] flexibly to communicate with humans.[10] Kanzi is a sentient, directed-intentional agent with a robust capacity to learn new behaviors. As a full grown male bonobo, he often has energy to burn. After a morning of word play in the research center, he is restless and twitchy. There are many games that he likes to play, but one of his favorites is <*ball*>. Imagine Kanzi as he paces about the room. He has learned the significatory relation between the <*ball*> lexigram and ball-playing and now, as his restlessness dominates and he becomes tuned to the possibilities for releasing his tension, the lexigram for <*ball*> becomes a ready-to-hand tool. He picks it out from the sheets of lexigrams that lie about the room, points at it, and looks around expectantly. When someone responds, perhaps by looking in his direction, perhaps by saying, "What would you like to do Kanzi?" the situation shifts, and new possibilities open up. Kanzi points again to the lexigram and then to his interlocutor. As with S, there is no need to suppose that Kanzi's lexigram use is an outward sign of some inner representational life, that Kanzi already has "in mind" the desire to play and merely uses the lexigrams to communicate this desire. On the externalist view being developed here, there are no *inner representations* to be communicated at all. Kanzi is drawn, according to myriad attentional pushes and pulls that are continually shaping, dissolving, and reshaping, to whatever is ready-to-hand in his environment, itself a product of Kanzi's past learning experiences and what is currently winning out in the competition for attention. Here, it is restlessness and excessive energy that compel Kanzi to reach for the <*ball*> lexigram. There is no plan here, no thought, "Now I have to find someone to play ball *with*"; rather, because balls are always thrown *to* someone or caught *from* someone, when Kanzi sees a person, he sees a potential playmate.

Now consider that signs such as the <*ball*> lexigram are tools, cognitive-enhancing tools to be sure, but, more basically, tools for effecting change in one's environment: by yelling "fire," I can induce panic in a group of people; by making a request, I can acquire food. Seen in this light, they are new additions to what is ready-to-hand. As one gains facility with these tools, as they become increasingly ready-to-hand, the more they open up the space of possibilities. With them one can plan—as Kanzi does when he sets up the conditions for ball-playing—and refer back to the past—as Kanzi does when he points to a bite on his arm and signs "Matata bite." As soon as Kanzi signs "you ball me," the utterance is now a thing in the world. It is a situation to which we can react. One could nod in assent, one could begin kicking the ball, or one could describe the situation: "*You* say, 'you ball me.'" Describing is a type of action, something we *do*. Our language-rich environment, overflowing with signs natural and artificial, presents an enormous pull to engage in this descriptive activity. Indeed, though they are not always visible—signs are not something we *see* unless they are words on a page or a billboard—the environment of a literate, adult human contains words more than it contains anything else. This is quite remarkable, particularly so when we consider how unaware we are of this singular fact.

Once S knows how to use a word, the word has become an object for S, an item of significance-for that triggers in S a set of responses. As words also function to bring what is not here and now to the fore, they can be used to extend our possibilities for action, for populating our world with new objects to which we can be drawn. When Fred, planning for his trip to Toronto, considers the various travel options available to him, he uses the representation tools that he is skilled with to lay the choices out before him. For Fred, this means running through the options propositionally in his head: "I could take the train, but that means I'd be coming back later than I'd like to; I could drive, but then I'd need to leave early to avoid traffic." For others, this might mean using a pen and paper to create a pro and con list for each choice. For those who prefer visual representations, using a mapping tool to compare travel routes and times might be best.

Which representation tools we prefer will likely be the ones we are most skilled with, a predilection that leads to a further deepening of the skill itself with each use. The more skillful we become in some activity, the more natural the exercise of the skill feels: indeed, for a master artist, the medium of their art—paint, music, words, dance—is so enmeshed in their embodiment that their emotions suffuse their art unbidden. So too, once one has become a sophisticated language user, the *feeling* that speaking and thinking is to trade in content grows: sentences seem to communicate ideas and intentions that are already *there*. But this sense that there is a *what* that we are communicating is analogous to the sense we have when we visually perceive a scene that perception is a smooth, consistent, and non-gappy process when it isn't (O'Regan and Noë 2001: 946). In both cases, the illusion is a consequence of our confidence in our skills. We know how to subtly shift and move to fill in visual details. Word-use is effortless and as the expectations they trigger are fulfilled by the unfolding situation, we feel in control of the entire experience; we understand what is going on. The more connections we have learned between words and possible responses, the more in line our expectations will be with the way that conversations and thoughts unfold and the more profound will be our sense that we understand *what* is being said. Add to this the self-referential possibilities that emerge out of word tools such as indexicals and it's easy to see how a solid sense of "I am thinking *this*" can develop. The indexical seems to cash out into some bit of content, something that I am aware of at a deeper, non-linguistic level. But "this" is just another word that I know how to use, and the deeper level to which it points is just another situation, one in which I am talking to myself—silently—using the representational tools I am so skilled with.

Consider Fred, admiring a red peony in his garden. In bare perception, he simply perceives the flower in an ongoing interaction that lasts as long as his attention is entrained. Imagine that during these moments of visual attention, Fred begins to feel joy and perhaps also becomes aware of some fleeting wisps of past flower experiences. There are no t-representations of the flower here, nothing in addition to the flower in front of him, his interaction with it, and the ongoing multilevel perceptual activity within Fred.[11] But now Sally sidles alongside and says, "I wonder what that red is like for you. For me it is so firey!" According to internalist accounts, Sally has just used some words to describe *her perception of the peony*, as though there is something here over and above her ongoing experience. Not so. Just as with Fred, Sally's perception of the peony is an ongoing process that the multilevel organism we call Sally is engaged in. Eliminativists (Churchland 1985; Dennett 2001) are right here that qualia *talk* does not entail qualia existence, but wrong to reject (if indeed they do, it is not always clear) subjective experience, a sentient creature's personal level responses to its world.[12]

On the other hand, to use words is to do something, and this activity is itself a part of the ongoing situation within which we are always embedded. Since we use words to describe situations, we can also use words to describe word-use itself. This is what Sally is doing here, using words to describe the peony, presenting this description to Fred, and asking him for his own description of the peony so that they can compare their *descriptions*. Sharing and comparing descriptions brings people together, and sometimes sets them apart, because they open a window into another person's learning history, the experiences they must have had for this object to elicit such responses.

I hear, "Close the door!" and immediately I am responding: my body turns and the door comes into view; I see the light filtering in between the door and the frame; I realize that it is ajar; I see my arm extending toward the doorknob; I pull it toward me both hearing and feeling the click as the latch catches. I do not internally represent <*door*> and a series of possible actions I might perform with doors and then infer what is required of me here; rather, I feel that I know what to do here.

Conclusion

Intentional agents direct their own behaviour; representationally intentional agents can represent these "intentions," to themselves and to others. Fred opens the refrigerator and fixes himself a salad lunch *even though* the leftover pizza beckons to him from the main shelf. He refrains from eating the pizza because he is trying to lose weight, most immediately in anticipation of his upcoming holiday in Mexico, but more generally because he prefers not to be overweight. On reductive accounts, such reasons have no teeth since it is the myriad subpersonal processes that are causally responsible for Fred's actions.

But once we abandon internal contents, we need another way of understanding how agents perform actions for distal reasons. A comprehensive account of Fred's behavior will need to explain what inhibited the pizza-eating reflex, and this in turn will involve an account of how non-occurrent factors played a causal role in his behavior.

Externalism offers such an account. Reasons play a causal role in human behavior by way of the representational tools we have developed. With words, we represent to ourselves and to others possible states of affair. As objects in the world to which we have developed responses, words have efficient causal force; as representational tools, however, they point beyond themselves, evoking responses to what is signified rather than to themselves. When a sentient agent develops a capacity for using representational tools, first in the very rudimentary way in which animals use distress calls to warn of danger, and then, in more sophisticated ways, using increasingly general and arbitrary symbols to perform increasingly abstract actions, the beginning of representational activity emerges. The dynamic interchange between representational tools that are easily produced and combined with the inclination to use them underwrites an explosive proliferation of words and word use. The world into which new humans are born is a language-rich one. A child's development into a literate adult is thus a symbiosis between its own learning capacities and the word-filled landscape in which it grows.

This view is parsimonious with evolutionary theory since it postulates only evolutionarily plausible differences to explain the wide cognitive gap between humans and nonhuman animals: flexible learning and tool-use capacities, a mechanism of expectation^{sys}, and a highly social nature. Simple representation-use, transmitted vertically through generations and laterally through communities, spreads rapidly, and as it develops into increasingly systematic structures, these in turn support and fuel further learning, use, and development. The technological snowball effect of this fortuitous combination yields the linguistic cognitive niche that the modern human develops within. This vast, rich, representation resource, then, the linguistic cognitive niche of the modern human, is what ultimately explains the cognitive gap between linguistically sophisticated humans and our closest ancestors.

Part III

Implications

9

Implications of Externalism

The next step for Externalism is to empirically test its hypotheses. If it's true that our proficiency with representational tools enables and deepens our representational intentionality—that it is a necessary condition of it—then when representation-use skill is low, we should see a corresponding deficit in deliberative capacity. As there is a wide range of representation tool modalities—verbal, visual, audial—there should be a respectively wide range of proficiencies and, perhaps, different possibilities for *what* is represented: with music, for example, a simple series of notes or an entire composition can call up a large swathe of fluid human experience. There is much work to be done to investigate the representational uses these different modalities can serve and the competencies that result from mastery of them. In the field of literacy studies, there is already a growing body of work that shows a clear link between the degree of language exposure in early childhood and cognitive development: a paucity of exposure leads to cognitive deficits and rich, robust linguistic experiences lead to cognitive advantages (Ardila et al. 2010; Bialystok 2001; Bramao et al. 2007; Castro-Caldas et al. 1999; Jensen, Burleson, and Sadauskas 2012; McNamee et al. 1985; Morais and Kolinsky 2005; Reis and Castro-Caldas 1997; Reis, Guerreiro, and Castro-Caldas 1994).

The other substantive claim of Externalism, that the expectation^{sys} capacity that makes representation-use possible is a general-purpose mechanism that did not evolve solely to support linguistic activity, is already supported by a surge of studies on expectation^{sys} effects ranging from physiological placebo/nocebo responses to medical interventions to cognitive deficits resulting from negative stereotyping. In the following chapter, I look at some of these in detail.

In Chapter 11, I offer an externalist assessment of our most promising AI systems. There is much misinformation and misunderstanding surrounding these increasingly prevalent systems, and we need to correct that if we are to use them wisely. Externalism gives us some clear metrics for determining the extent to which a given system is a representation-using one or merely a representation-isomorphic one. When only the latter is the case, the degree to which language can be learned will be limited in ways that we have yet to understand fully. Work in the field of Emergent Communication[1] will help with this.

Finally, in Chapter 12, I look at some of the philosophical implications of Externalism. As analytic philosophers work almost exclusively within an internalist, representationalist paradigm, these are rather wide-ranging; I look at only those that seem most pressing.

10

Empirical Implications

In this chapter, I review recent work in the cognitive sciences that examines the causes and effects of expectations on our cognitive behavior as well as on physiological responses more generally. I will begin with an overview of some of the areas being studied and then select a few to look at in more detail. Finally, I will look at how well these findings do and don't support the externalist picture of intentionality being developed here.

The literature makes an important distinction between *implicit* expectations—the neural priming consequences of our synaptic pathways—and *explicit* expectations—the consequences of our beliefs and concepts. The former are factors of which an agent cannot become introspectively aware, for example, activation potentials of neurons in one's brain that encode expectations about word order; the latter are factors of which an agent *can* become aware, for example, beliefs about appropriate attire for different social contexts. Implicit expectations manifest as neural activity, while explicit expectations manifest as episodic flashes, trains of thought, feelings, emotions, and behavior. As before, I will adopt the convention of applying the subscript "sub" to "expectation" to denote implicit or *sub*system $\text{expectations}_{\text{sub}}$ and the superscript "sys" to denote explicit or *system* $\text{expectations}^{\text{sys}}$; alone, the term denotes expectations of the system whole with no distinction between the scalar aspects of its constituent features.

Research into the cognitive role of $\text{expectations}^{\text{sys}}$ divides into two complementary approaches: (1) those that study the subsystem attributes of subjects primed with $\text{expectations}^{\text{sys}}$ by comparing their neural processes with those of controls not so primed; and (2) those that study the system attributes of subjects primed with $\text{expectations}^{\text{sys}}$ by comparing their behavior, including self-reports and other measurable system attributes, for example, rates of respiration and perspiration, with those of controls not so primed. In the first sort of experiment, researchers look for correlations between subsystem activity and system changes to identify the brain areas responsible for those system processes. In terms of a mechanistic framework, they are trying to uncover the subsystem components of expectation^{sys} mechanisms. With respect to fleshing out our map of the brain, they are identifying the neurological differences between different *kinds* of expectation^{sys}, for example, positively or negatively valanced ones. This is important work. But it is the second type of study—identification of *system* components—I am interested in here: if we want to better understand the role that expectation^{sys} plays in deliberative cognition, we need to examine the system attributes[1] that are causes and effects of it. Knowledge of this will in turn help us direct the neural investigations more fruitfully as well.

The three main areas of expectation^{sys} research are in placebo/nocebo effects of medical interventions, stereotype effects on cognitive performance, and perception effects. Studies in these areas uncover what expectation^{sys} effects exist, investigate how the expectation^{sys} mechanism yields these effects, and test methods for directing, augmenting or, in the case of negative stereotypes, subverting expectation^{sys} effects. Note that subjects are not always aware of their own expectations^{sys}, that is, awareness *of* expectation is not a necessary condition of expectation^{sys}: blindsight patients, for example, have expectations^{sys} about objects in their visual field but report no awareness of this.[2] Indeed, the burgeoning area of expectation^{sys} bias research illustrates just how pervasive and yet unnoticed our expectations^{sys} can be.

Expectations^{sys} can enhance or impair memory, influence decisions, inform beliefs both rational and irrational, influence our capacity to learn, motivate action, motivate skill development, motivate learning, and motivate attention.[3] Positive or negative beliefs about the efficacy of some medical intervention (placebo/nocebo effects) have been shown to have, respectively, positive or negative effects on both the subject's perception of the outcome, that is, how well they judge the treatment to have worked,[4] as well as on the outcome itself, that is, whether the underlying condition changed.[5] Analogously, positive and negative beliefs about the performance capacities of some groups (stereotype effects) yield expectation^{sys} biases that influence both behavior with respect to these groups, for example, a stereotype bias might lead one to treat members of a gender group preferentially, and performance on specific tasks, for example, members of a negatively stereotyped group with respect to some ability will exhibit diminished performance in stereotype contexts.[6] Finally, expectations^{sys} about the details of occurrent situations, including both local, physical conditions as well as broader, social ones, influence perception, for example, a facial expression might be perceived as angry in one condition and fearful in another.[7] A sampling of some studies will help ground our discussion.

In one study (Assefi and Garry 2003), half the subjects were given a drink that they were told, falsely, contained vodka. All subjects were directed to perform a memory task involving misleading information. The half who thought they were drinking vodka were more susceptible to the misleading information and consequently performed worse than the control. In another study using the same memory task (Clifasefi et al. 2007), test subjects were given a drug that they were told was used by the military to enhance soldiers' performance. This group was less susceptible to the misleading information and consequently performed better than the control. "In other words, within the same task, the opposite expectancies produced the opposite results" (Michael, Garry, and Kirsch 2012: 151).

In another study (Colagiuri, Livesey, and Harris 2011), placebo expectations were manipulated to see whether they had cognitive effects. Prior to a response/discrimination task—finding a rotated letter *T* on a screen filled with distracting figures—subjects were directed to sniff a cotton pad, unscented for half of the subjects, the control, and scented for the other half, the test group. Test subjects were then further subdivided into three groups: one was told that the aroma would enhance their performance, one was told it would diminish their performance, and the final group was told it would have no effect on their performance. The results followed the expectations^{sys} set by the experiment: those who expected^{sys} their performance to improve found the *T* faster, those who expected^{sys} their performance to be diminished took longer to find the *T*, and those who expected^{sys} no effect performed on par with the control.

Marketing companies have been using expectation^{sys} effects to induce product preferences for over a century. This is hardly surprising; it works. Many studies demonstrate that when people expect^{sys} a quality product (and years of branding establish these expectations^{sys}), they prefer those products to others even when there is no difference between the products themselves. In one study (Wilcox, Roggeveen, and Grewal 2011), subjects were asked to taste a piece of chocolate and rate it. Test subjects were divided into four groups:

those told before tasting either that the chocolate was from Switzerland or from China and those told after tasting either that the chocolate was from Switzerland or from China. Those in the two after-tasting groups showed no effects, but those told prior to tasting that the chocolate was from Switzerland liked the chocolate more than did those told prior to tasting that the chocolate was from China. The Swiss are known for their quality chocolate and this background knowledge seems to have exerted expectation^{sys} effects on the subjective reports.

I turn now to a more detailed look at three studies investigating perception, stereotype, and placebo expectation^{sys} effects respectively.

Perception Effects

I've chosen to examine this study (Diekhof et al. 2011) because it investigates the role that expectations^{sys} play in developing t-representations. Specifically, the study manipulates subjects' expectations^{sys} for facial expressions to invoke different emotional responses to presented facial expressions. In addition to showing that mental imagery directed by expectation^{sys} yields an emotion-modulating effect that expectation^{sys} alone does not, the study demonstrates that there are significant effect differences between mental imagery that is explicitly paired with expectations^{sys}—mental representations on Externalism—and mental imagery that arises incidentally with expectations^{sys}—*not* mental representations on Externalism.

Subjects were asked to rate presented human facial expressions as more or less fearful after a preparatory anticipation phase. In the expectation phase, subjects were presented with one of three cues that indicated the degree of fearfulness of the face to follow—neutral, mildly, or highly fearful. Occasionally, the cue was misleading—it indicated a less fearful face than the one to follow—but subjects were unaware of this deception. Test subjects, members of the imagery group, were directed to mentally imagine the facial expression to follow according to the anticipation cue, e.g., if a highly fearful face was cued, they were to conjure a mental image of a highly fearful face. Control subjects, members of the no-imagery group, were simply told that the cue predicted the image to follow. For both groups, whenever a misleading cue was presented during the anticipation phase, it was always less fearful than the face to follow. The question of interest here was whether the response to the stimulus was modulated by either the expectation^{sys} of reduced fearfulness alone or by the explicit pairing of mental imagery with the expectation^{sys} of reduced fearfulness. If the latter, then modulating or preparing for emotional responses is one plausible role of representational mental imagery[8] and the expectation^{sys} mechanism that is part and parcel of that representational system.

Results were interesting. Test subjects in the imagery group "rated the highly fearful expression as significantly less fearful if preceded by the misleading cue ($t=-2.32$, $p=0.034$) and as significantly more fearful if preceded by the correct cue ($t=2.57$, $p=0.021$), when compared to the group that was not instructed to imagine the expression during anticipation" (Diekhof et al. 2011: 1707). That this difference might have been influenced by discrepancies between subjects' sensitivity to fearful faces was ruled out by a "'pure' rating phase at the end of the scanning period" (Diekhof et al. 2011: 1708) during which fearful facial expressions were presented without cues. There was no significant difference between group responses on this task. A corresponding finding was made at the neural level: test subjects anticipating fearful expressions showed a marked "increase in activation of the left FG when compared to imagination of neutral facial displays and to activation in the same comparison in the NoImagery group" (Diekhof et al. 2011: 1708).

When expectations^{sys} are explicitly paired with mental imagery, when a particular situation is *actively* imagined, that active imagining is a case of representation-use, namely, a continual re-creating of the mental image that is driven by the expectation^{sys} for the thing imagined. When there are only wispy episodic mental flashes, namely expectations^{sys} but no use of them as representations of what is to come—this is the NoImagery case—there is no marked change in subsequent behavior. Representation-use has cognitive effects. Some of these are merely continuum effects: as this study demonstrates, expectation$_{sys}$ of a less fearful face, even without mental imaging, will modulate emotional response to some degree:

> The NoImagery group showed a nonsignificant, but nevertheless positive correlation between anticipatory activation in the FG . . . These results suggest that an anticipatory engagement of the FG, especially during mental imagery of a mildly compared to a highly fearful expression, indeed biased subsequent perceptual judgment and reduced the subjectively perceived

> fear intensity in the misleading condition. This finding conforms well with several previous observations (e.g., Ploner et al. 2010) and supports the assumption that anticipation involves crude (subconscious) forms of mental simulation. (Diekhof et al. 2011: 1708)

Mental representations here function to augment the modulatory emotional response by serving as an anchor for attention. In the NoImagery group, expectation^{sys} has some modulatory effect, but it is mild because there are so many other processes drawing the attentional resources of the overall system. By creating a mental image of the expected face, the Imagery group effectively anchored their attention to what was expected. Representations yield these strengthening effects because expectations^{sys} are normally vague, amorphous feelings accompanied by associative episodic flashes that are continually changing and of which subjects are only intermittently aware. By collecting these into a single structure—a label or an image—the subject creates a stable target for attention to latch onto. Since in imagining a perception we are effectively simulating the act of perceiving itself (Bermpohl et al. 2006; Boly et al. 2007; Carlsson et al. 2000; Porro et al. 2002; Nitschke et al. 2006), actively imagining a less fearful face lowers the emotional stress response of the subject. When the actual face is presented, unless there is a huge mismatch with what was expected, there will be no draw on resources to get more sensory details. In essence, the subject will perceive the anticipated face rather than the actual one. Representations thus serve to focus the attention by providing stable anchors for the attention. This is a cognitive advantage of degree since attention will be drawn subconsciously here and there intermittently anyway.

Representations also provide new *kinds* of cognitive advantages. Subconscious draws on attention are always to features of the occurrent situation, internal and external, as well as to short-term expectations triggered by responses to these same features. Representations, as we've just seen, are a tool for stabilizing attention on one of these features. But as multilevel intentional systems, we are continually exchanging resources with our environment in an ongoing interactive, responsive cycle. Representations, by standing in for some situation or thing, bring to the occurrent situation what is not physically present; they broaden the scope of a system's environment. They make responses to spatiotemporally distant or even unreal situations possible. This ability to expand one's world beyond the here and now via representations is both a boon and a burden, as the next study demonstrates.

Stereotype Effects

Stereotype effects are changes in system behavior caused by a belief that one has or does not have a capacity of some sort. Typically, such beliefs are linked to group membership—for example, *females perform worse than males in analytic tasks*—but they need not be—for example, *I am bad at math*. Interestingly, a subject does not need to hold the belief themselves for the effect to occur; it is enough that they believe that others have the belief and thus feel pressure to demonstrate that the belief is unfounded. Some studies (Schmader and Johns 2003; Croizet et al. 2004; Inzlicht, McKay, and Aronson 2006) argue that stereotype expectations^{sys} induce anxiety, which in turn overloads working memory resources. This overall depletion of cognitive resources explains the negative impact on performance. But anxiety is a large-scale, systemic response that can be caused by many more factors besides negative stereotype expectations. One study (Johns, Inzlicht, and Schmader 2008), sought to isolate the specific psychological processes caused by stereotype expectations that negatively impacted performance. They reasoned that if they could manipulate subjects' beliefs about the origin of the psychological processes inhibiting their performance—disassociating them from the negative stereotypes—they could mitigate the negative effects. If successful, this would have the dual outcome of identifying a system-level consequence of expectations^{sys} and a constructive intervention for mitigating the effects of stereotype bias. I will summarize the four experiments, going into detail where there are some important findings, but generally presenting the results of the four taken together.

The key insight driving the study as a whole is that, contrary to previous hypotheses, the anxiety induced by negative stereotype situations is not the relevant factor in decreased performance; rather, the efforts of subjects to control and suppress their anxiety overloads executive functions which, in turn, leads to a decrease in performance.

Study 1

All subjects were women who were aware of a negative stereotype around women and competence in mathematics. In all conditions, subjects were seated three to a row, side by side, facing computers. In the *threat* condition, the test subject was flanked by males (bogus subjects) on either side and was told, by a male experimenter, that the purpose of the study was to assess whether there are mathematical aptitude differences between males and females. In the *no-threat* condition, three female subjects were seated side by side and were told, by a female experimenter, that the purpose of the study was to assess the problem-solving aptitudes of college students. All subjects were told that the math test/problem-solving task would be divided into two parts interspersed with "filler" tasks, an anxiety and working memory test, which were, in actuality, the measures of greatest interest. The anxiety test was a dot probe task designed around the assumption that anxiety increases attention toward threat. Subjects were asked to track dots associated with either a neutral or an anxiety-related word and response times were measured. Higher scores indicated attention to anxiety-related words while lower scores indicated no special attention. If subjects were actively trying to suppress their anxiety, they would avoid looking at those words and hence have a lower score than those not actively suppressing. The dot probe task was thus used both as a baseline measure of anxiety and as a measure of the degree to which anxiety was being actively suppressed. The working memory test was a reading span task (Schmader and Johns 2003) in which subjects were first directed to memorize a set of words, then to count the vowels in a series of sentences, and then to recall as many of the memorized words as possible.

To measure the spontaneous suppression of anxiety as a system-level phenomenon, the researchers manipulated subjects' explicit knowledge of key aspects of the experiment. Thus, subjects were further divided into an *implicit* condition group, whose members were told that reaction time was being measured but not why it was being measured, and an *explicit* condition group, whose members were told both that reaction time was being measured and that the measure was being used to assess anxiety. There were, therefore, four groups in total: *implicit threat, implicit no-threat, explicit threat, explicit no-threat.* Researchers expected that the *implicit threat* group would engage in unconscious anxiety suppression behavior, while those in the *explicit threat* group would use system-level behavior to actively suppress their anxiety, for example, by keeping their attention focused on non-threatening aspects of the situation.

As predicted, women in the *threat* condition did poorly on the working memory test, remembering fewer words than those in the *no-threat* condition. This result confirmed that stereotype threat does indeed overload executive functioning. With respect to anxiety suppression, subjects in the *implicit threat* condition had higher scores on the dot probe test than those in the *explicit threat* condition. This suggests that while the attention of subjects in the *explicit threat* condition was likely drawn to the anxiety-related words, they were actively redirecting their attention away from those words. The two *no-threat* condition groups showed no significant difference in attention toward anxiety-related words. This confirms that the *threat* condition heightened anxiety and that the *explicit* condition created a motivation for suppressing anxiety.

Study 2

In this study, the goal was to directly manipulate subjects' emotional regulation to study the effects of emotion suppression on working memory. There were three test conditions: in the *threat-only* condition, subjects were given no instructions beyond those describing the mathematical aptitude tests; in the *suppression* condition, subjects "were instructed to suppress and hide all feelings and emotions they had while preparing for, thinking about, and taking the test, and to 'behave in such a way that a person watching you would not know you are feeling anything at all'"; and, in the critical *reappraisal* condition, subjects "were instructed to look at the test in an objective manner, to adopt a neutral attitude as they prepared for, thought about, and took the test, and to 'think about [the test] objectively and analytically rather than as personally, or in any way, emotionally relevant to you'" (Johns, Inzlicht, and Schmader 2008: 9). Working memory was tested using a Stroop test.

Subjects in the *reappraisal* condition made fewer errors on the Stroop test than those in the other two conditions, indicating that working memory was less overloaded for them. Likewise, they made fewer errors on the math aptitude test than subjects in the other two conditions, indicating that when subjects were directed to reappraise the situation in a way that led them to stop suppression, the negative stereotype effect was eliminated.

Study 3

In this study, the *reappraisal* condition was tested against a *threat-only* and *non-threat* condition. As in Study 2, the negative stereotype effect was eliminated for the *reappraisal* group: their results on both the working memory and mathematical aptitude tests were the same as those of the subjects in the *non-threat* condition. Those in the *threat* condition who were not directed to reappraise their responses performed poorly in both. Interestingly, subjects across the two threat conditions reported feeling anxious, but those in the *reappraisal* group were able to effectively inhibit their own suppression tendencies and this, in turn, resulted in better performance.

Study 4

Here, the researchers used a different stereotyped group to generalize the results, which proved to be consistent with the first three studies.

The researchers conclude that “results from the four experiments develop a chain of causality to provide converging evidence that stereotype threat depletes executive resources via emotion focused regulation” (Spencer et al. 2005: 17). This “maladaptive form of emotion regulation” they argue is the specific cause of diminished performance. Furthermore, a system-level intervention of *reappraisal* effectively restores performance, not by eliminating anxiety, but by eliminating anxiety control. Thus, it is an agent’s expectation^{sys} of a threat condition that instigates an anxiety response, but it is the expectation^{sys} that *this very anxiety will undermine performance* that motivates subjects to explicitly or implicitly try to suppress their anxiety response. A causal chain of expectations^{sys} are at work here: expectation^{sys} of threat triggers an anxiety response and awareness of anxiety triggers further anxiety, which leads to poor performance. This meta-anxiety, we might call it, motivates the subsequent anxiety suppression behavior. But it is this very act of suppression, ironically, that undermines the performance they are working so hard to optimize. The effective correction is emotional distancing through detachment, represented here by the *reappraisal* condition: since the stereotype condition depends on identification with a group, actively separating self-identity from it alters the entire situation so that the stereotype condition no longer exists.

In terms of expectation, *suppression* is an ineffective strategy because it targets only the effect of threat expectation^{sys}, namely anxiety; *reappraisal* is effective because it targets the *cause* of threat expectation^{sys}, namely, the stereotype condition. Once this expectation^{sys} is eliminated, there is no anxiety with which to contend.

Placebo Effects

One clear finding of the stereotype expectation research here is that expectations^{sys} play a causal role in our subjective appraisals of our own performances and the situations we find ourselves in. This can have far-reaching consequences since negative expectations^{sys}, for example, of threat, can lead us to feel anxious, and anxiety and the impulse to control anxiety impact a wide range of behavior including cognitive performance. This is an important finding. But perhaps more intriguing are the expectation^{sys} effects in areas we assume are immune to subjective appraisal.

In a paper (Colloca and Benedetti 2005) arguing that expectation^{sys} effects undermine even the most rigorous clinical trials—a finding that has immense consequences for the entire corpus of pharmaceutical research—the authors reviewed clinical trials that controlled for expectation^{sys} effects by treating patients without their knowledge. The drugs in these studies were "administered through hidden infusions by machines[33,35,55–58] (Colloca and Benedetti 2005: 548). Hidden administration of pain relief drugs—no-expectation^{sys} condition—was less effective than both placebo and non-placebo transparent administration of drugs—expectation^{sys} condition. Where some positive effect did occur in the no-expectation^{sys} condition, it was significantly less than in the expectation conditions. In another study (Matheson and Cole 2004) that investigated the "effect of methylphenidate on glucose metabolism in the brain" (Colloca and Benedetti 2005: 550), it was found that the effect in the brain was 50 percent greater in cocaine abusers who expected treatment than in those who were administered the treatment without their knowledge. These findings, the authors urge, give us good reason for focusing more research on uncovering the ways in which expectations^{sys} play a role in medical interventions as "we have no a priori knowledge of which substances act on pain pathways and which on expectation mechanisms—and almost all drugs might interfere with the top-down mechanisms" (Colloca and Benedetti 2005: 546).

Conclusion

Expectation is a key mechanism of the entire prediction processing system. The bulk of the empirical work in this area is in identifying the neural expectation pathways that correlate with expectation^{sys} so that we can better understand how expectations^{sys} yield substantive cognitive effects as well as medical ones, not just in the domain of subjective appraisals of pain and wellness but also in objective measures of disease.

We must be careful how we think about the component processes of this mechanism. Causally, expectations^{sys} can affect only macrolevel behavior; however, as those effects are themselves constituted by microprocesses, there is an indirect causal link to them. Dramatic subsystem effects such as cellular repair will likely require broad subsystem changes. Perhaps one way that expectations^{sys} can yield these is via attention^{sys}. When a subject strongly expects^{sys} some change, they are actively attending to their expectations^{sys}, that is, using representations to maintain their focus. An action in which representations are continually being invoked in order to bring attention back to an anchor of focus is constituted by entrained activity at multiple scalar levels. Strong expectations^{sys} can help to motivate and maintain such entrainment, which, in turn, entrenches it further, spreading its influence until more and more of the entire system is engaged in the activity. This widespread activity is what ultimately leads to cellular change.

As these studies demonstrate, it is difficult to intervene in the causal chain of subsystem activities once expectations^{sys} have become entrenched. Intervention in their development to begin with, however, is an effective way of directing expectations to support success, even when that involves physiological effects such as healing. This can be achieved in the medical context through frequent patient-doctor consultations and more whole-person care generally. Negative expectations^{sys} can be unlearned too, through explicit awareness and reassociation training. Becoming aware of the biases that govern our thinking is the critical first step in undoing their negative effects. Increasing the diversity of one's experiences, through reading and exposure to new situations, will eventually replace them with more nuanced expectations. Thus while expectations^{sys} enable representation-use in the first place, the development of the latter in turn manages and directs further expectations^{sys}. Such directed, stabilized, or moderated expectations^{sys}, in their turn, open up new possibilities for action. It is this virtuous circle of expectations on the one hand and the development of the representation-use skill on the other that characterizes the representational intentional agent

11

Technological Implications

The central question of this chapter is whether our best AI systems are or can become representation-users at all. In the context of the current AI frenzy in popular culture, this is a critical question. If having a deliberative capacity is a necessary condition for making the kind of judgments required in driving, prescribing medication, performing surgery, or determining penal sentences—domains in which AI systems are poised to step in—then we'd better know how to assess whether they have that capacity or not. Here I begin this work by using Externalism to give an analysis of our most promising candidates, autonomous robots that can act without external control.

Viewed through a wide lens, there are two distinct approaches to creating autonomous systems: top-down models that use cognitive systems to control behavior and bottom-up ones that are guided by structure and simple, low-level, sensorimotor loops.[1] Today, many autonomous robots utilize a hybrid approach. The Roomba® is a good example of this: for the most part, its various sensors trigger local motor responses that ensure the entire system avoids obstacles. For instance, activation of a bumper sensor triggers a motor response to move backwards. But it also has a floor-mapping system that uses data from various cameras to produce a map of the area to be cleaned, which is used by certain control subsystems to ensure that the entire space is eventually covered. Such cross-utilization of ideas is certainly promising, but, theoretically, these two distinct approaches still mark a clear division in the field of AI research today. The bulk of the attention and money goes to research into top-down, data-driven approaches, for example, Deep Learning Neural Networks (DNNs), virtual systems that can easily be integrated into a wide range of technologies, including the search engines and chatbots that are used daily by many. Work in bottom-up robotics, sometimes called

"Biological" or "Enactivist" AI[2], is not as well known, gets less funding, and consequently has fewer researchers working in the area. This is a shame as there is much important work being done here. After an externalist analysis of the top-down approaches, I will end with a word about why incorporating bottom-up insights is essential to eventual success.

Autonomous Driving Assistance Systems (ADASs) provide a good example of advanced autonomous robotics. Currently, there are no vehicles that can drive completely autonomously[3] and there are serious safety concerns surrounding the use of these systems on public motorways. When we consider the many factors a driver must attend to and then use in making good judgments, we can see how difficult the task of creating a completely autonomous ADAS is. If I see that up ahead a youth is riding on a skateboard near the side of the road, I will become extra alert, hover my foot over the brake pedal, and swerve into the next lane if there is room. This quick practical judgment is grounded in my robust understanding of a non-finite (and hence non-quantifiable) set of possibly relevant factors—the skill and age of the rider, the likelihood of falling into the road at that speed, whether there is a social trend for risky skateboarding, and so on. Deciding to apply the brakes because I am too close to the car ahead, on the other hand, does not require much in the way of judgment since the situation is relatively constrained and lends itself to a quantifiable response: determine whether the distance between the car and the one in front is below a certain threshold for the current road conditions. The question of Level 5 ADAS safety, then, ultimately rests on whether and to what degree such a system could ever make non-quantifiable practical judgments of the sort that human drivers regularly make. To make this assessment, we need to look more closely at how they work.

ADASs are built according to a functional schema that is very loosely inspired by the human system: its sensory systems (perception) gather data about the external environment, and its low-level (haptic) and high-level (deliberative) control systems use this information to orient and move the vehicle. In situations of uncertainty, when the vehicle is moving and the environment is continually changing, a key problem is maintaining an accurate model for the control systems. If the car has moved to the left to avoid a pothole in the road, for example, the control system calculations must be updated with this new baseline. Machine learning algorithms are useful in these situations because these techniques excel at extracting features from the environment for a specifiable task, in this case, determining the current vehicle model. Many different DNN heuristics—reinforcement learning as an optimization technique, for example—have improved the way in which these data-driven processes handle data, so we will need to look more closely at how they work before we can analyze the general approach.

Simple artificial neural networks can be thought of as functional models of neural activity. Numerical assignments of "nodes" represent the activation levels of individual neurons in the network, and the learned response relations between nodes, the synapses, are encoded as connection weights, positive or negative numbers that determine the strength of the connection between any two nodes. Network activity is modeled by calculating the activation level of each node according to an activation function, given an input, the current activation levels of each node, and the connection weights between them.

Networks are trained using a training set and an algorithm for adjusting the values of nodes and weights according to performance:[4] initially, the system's node activation values and connection weights might be randomly set; then, for each input and optimal output pair in the training set, the network is run and node and connection weights are very slightly adjusted to reduce the error between output and optimal output. A training set might consist of 500 digital photographs of cats and dogs, input as 300×300 image pixel vectors, each of which is paired with an ideal output, activation of one of two nodes labeled "cat" and "dog" respectively. With many thousands of iterations of such training, once the error between the system's output and the optimal output has been reduced to near 0, the network is judged as trained.

Of course, whether or not a network is good at whatever it has been trained for must be determined by testing it on novel inputs. Simple networks might be adept in ideal situations, for example, when an input image is simple, clear, and in the same orientation as testing images. But when there is a great deal of noise, for example, when the lighting is different from that in the training set or there are many kinds of objects in an image, more complex networks are needed. Convolutional neural networks (CNNs), which are a kind of DNN, are a good choice since they excel at learning very fine-grained object discriminations even in very noisy situations.

In contrast with simple networks, DNNs have multiple layers (sometimes hundreds) of different types of nodes (e.g., pooling, convolutional, nonlinear) that work both in sequence and in parallel with or without recursive feedback looping to create incredibly nuanced, fine-grained discriminations. They are especially useful in situations in which the inputs are unknown or need to be gathered "on the fly" from the environment: networks are stacked together, hence the moniker "deep," with some first weeding through the initial data to reduce it to a set of potentially useful inputs that other networks, specialized for different subtasks, can process and then feedback to the initial weeding layers for more targeted data collection.

Since an ADAS control system needs an accurate, up-to-the-second car model to function, DNNs are an essential component of its sensory system. With these, Level 4 ADASs—no human control in a limited area and for limited driving—have purportedly been achieved ("Automated Driving Vehicle Approved" 2023; Ludlow 2024), although integrating them into the public sphere is proving challenging and dangerous (Cuthbertson 2024; Palmer 2024).

There is no question that ADAS technologies are marvels of engineering and extremely useful as Automated Driving *Assisted* Systems. But can they ever achieve full autonomous status, that is, make the requisite practical judgments that require attunement to the salient features of a situation? When we remember that these features are typically not directly available to sense detectors, e.g., that the skateboarder up ahead is *being reckless,* we can see what an enormous challenge this is; however, some progress toward answering the question can be made now by assessing their capacity for representation-use more generally.

The first thing to notice about these systems is that they implement only functional models of human sensory-decision-action systems, not structural ones. While the DNNs that make up the various components of the sensory-decision-action subsystems are composed of multiple layers of networks in ways that mirror human nervous systems, they are not structurally related to the system whole, in this case the car, in the way that the human nervous system is structurally related to its body. Why is this important?

One reason is timing. Different temporal relations hold between mechanisms and their components according to whether they are constitutive or causal components: causes precede their effects, but mechanisms and their constituents are active *cotemporally.*[5] The ostensible human decision center—the brain—does not first complete its computations and then output these to an action center, as is the case in the linear, causal sequence followed by the various control systems of an ADAS. Neural processing is never complete, not even for a given task, because the system is continually acting, working, and responding *at the same time* the neural processing is occurring. Indeed, and this is the second reason, because the motor system is itself a part of the nervous system, there are no neural processing "outputs" to be relayed to it. Certainly, nervous system activity is constituted by a series of input/output interactions, but these causal relations are themselves *constituents* of higher-level system activity. Motor neural activations cause muscle contractions and relaxations via neuromuscular junctions—this is intralevel causal activity. But because muscles are wrapped around bones, muscular contractions and relaxations push and pull bones. Because bones form the core of body parts, arms and legs move when bones move. Muscles don't *cause* bodies to move; rather, because of these structural relations, muscular activity *constitutes* body movement. By modeling only the functional relations between sense and control systems, constitutive structural relations are mistakenly treated as causal ones.

This error is a critical one. As we've seen, a system's perceptual recognition of what is significant-for it emerges out of the multilevel, coordinated effort to maintain its own systemic integrity. Perception mechanisms certainly include mechanisms for discriminating between objects, and DNN classifiers are getting better and better at identifying them. But the constraints that emerge out of a multilevel system's constituency relations dictate what needs classifying, not as a set of conditions that inhibit (or not) this or that discriminatory function, but as part and parcel of the discriminatory process itself. In other words, that a perception mechanism is situated within a system that moves through its environment in particular ways, that has these needs and goals, that is motivated to act in these ways, and so on, will dictate what is significant-for it. On unilevel functional DNN models, in contrast, significance-for is reduced to an implementation constraint, for example, a reward parameter, that is, once again, a causal component of the system. As such, it will always be behind and out of touch with the changing needs of the system whole.

In a nutshell, then, the key problem with the top-down approach used in current ADASs is the assumption that lies behind it, namely, that a functional model of something as structurally complex as a cognitive agent could ever *be* one:

> What proponents of this type of functional analysis fail to notice is that functional and structural properties of components are interdependent: both are necessary, mutually constraining aspects of a mechanistic explanation. On one hand, the functional properties of a black box constrain the range of structural components that can exhibit those functional properties. On the other hand, a set of structural components can only exhibit certain functional properties and not others. (Piccinini and Craver 2011: 301)

The devil really is in the details.

How might this lack of constituency relations between control and action systems affect the potential for representation-use? Let's imagine a robot (R) designed along the lines described here, outfitted with various perception devices—cameras, lidar, radar—and equipped with complex, multi-layered, data-driven sensing, decision, language processing, and control systems (S). Suppose that when the sensory system identifies an edge (US), the control system gives a command to move away (UR), to avoid US. Suppose further that the language system has learned a symbol for US, the label "edge alert" (CS), such that when the sensory system identifies the string "edge alert" and this is input to the control system, the command to move away (CR), to avoid the string source, is output *and* the sensory system is primed for edges. Now R's sensors detect the string "edge alert," and the control system moves R away while at the same time priming its sensory system for detection of edges.

A t-rep relation is certainly being instantiated here: R's sensory-control system S identifies the input "edge alert" (CS) and responds by moving R away (CR/UR) from the source *and* priming the sensory system—expectation$_{sub}$—for edge (US) detection. Neither edges nor "edge alert" strings are significant-for R or S; however, in virtue of S's capacity to identify and develop responses to edges, they are significant-in S.[6] Although significance-in relations cannot support representation-use behavior for all the reasons rehearsed in Chapters 4 and 7, we might call such t-rep relations *representation-isomorphic*: important representational relations between sign and signified are instantiated in the system, but there is no intentional use of this relation.

Models that are founded on representation-isomorphic relations will always face the intractable problem of determining which significant-in objectivist classes are significant-for—salient to—a given task or situation. But this has the process the wrong way round. Salience isn't a problem to be solved; it drives system behavior. The system probes its environment for significance-for, and this probing-perceiving process yields fluid situations (not just the objectivist categories that our words pick out), the boundaries of which are continually changing as an intentional agent acts. Because significant-in associations are meaningless *for* such systems, sometimes a very small variation in a situation, a change in an object's orientation, for example, will result in catastrophic misidentifications (Bowers et al. 2023; Nguyen et al. 2015). Adversarial attack (Cen et al. 2015) will always be a source of vulnerability for such brittle systems.

Much can be achieved with a representation-isomorphic capacity, as we are seeing today with the success of AI systems, but deliberative reasoning will not be among those accomplishments. Autonomous robots that model only the functional relations between sensing and acting will be lacking in the essential structural relations that integrate these processes into a system whole. As Carl Craver (2006) reminds us, "Tinker-toy models simulate human tic-tac-toe competence, but nobody believes that a blueprint of the model's grinding gears explains how human beings play" (355). I will end this chapter by looking at some possibilities for incorporating such structure into current autonomous systems.

Multilevel Systems and Bottom-Up AI

Many of the substructures of a vehicle are good examples of constitutive organization. A car's wheels, for example, are connected to one another by axles, which are connected to a central shaft on top of which the entire car body sits. These simple physical connections ensure that when one wheel begins to rotate, they all do, and the entire vehicle moves forward. In other words, the wheels do not cause the car to move forward; their movement constitutes it.

Implementing mechanical constitutive relations of this sort is something engineers already do exceedingly well. The puzzle in the case of autonomous systems such as ADASs is how to integrate the "cognitive" systems into the whole in a constitutive rather than merely causal way. What might a multilevel cognitive system look like? Current bottom-up methodologies inspired by evolution and enactivist insights give us a clue. On these approaches, system design is grounded in the idea that cognitive processes such as perception are continuous with life processes. Engineering is thus guided by two principles:

(1) Constitutive autonomy: Systems must be capable of generating their own systemic identity at some level of description;
(2) Adaptivity: Systems must have the capacity to actively regulate their ongoing sensorimotor interaction in relation to viability constraints (Froese and Ziemke 2009: 482).

The first principle is a recognition that action intentionality is fundamental: for any system to be autonomous, it must have intrinsic motivation. The second principle uses the notion of adaptivity to creep toward directed intentionality:[7] the more complex a system's sensorimotor interactions with its environment become, the more it directs itself toward the objects in its world. Soft robots that can change their shape (Shah et al. 2021) are an example of design according to these principles: when the terrain is flat or downhill, the robot takes on a cylindrical shape and rolls; when the slope angles up, the robot flattens and switches to a crawling gait.

Likewise, an ADAS might have integrated intelligent sensors, such as tires that change their shape in response to whether the terrain is bumpy or wet. Indeed, its control system would need to be integrated into every aspect of system mechanics. Perhaps a primary motivation for non-collision could be accomplished by a host of tight sensorimotor loops that together ensure that the system continually moves away from potential impacts. System movement would likely be irregular or twitchy, so another haptic, feedback system would need to be continually adjusting movements to keep them regular and smooth. Long-term system goals (e.g., *reach destination X*) could be integrated with the local "goals" of such subsystem activity via feedback loops that keep the system moving in a specified direction while maintaining optimal local conditions, that is, non-collision.

Notice what sorts of things would become significant-for such a system: terrain conditions, obstacles, speed, fuel levels. What would not be significant at all are the kinds of object categories that current ADAS systems are trained to classify: pedestrians, traffic lights, electronic bulletins. These are only significant-for beings for whom interaction with people, traveling from place to place, being a law-abiding citizen, and so on, is an integral part of its being. Why would such objects be significant-for a system that is motivated by non-collision and reaching destinations? As soon as we realize how entwined significance-for is with intentional motivation, we see that any fully autonomous ADAS that could be relied upon to make the safety judgments that a human might make would not only require a capacity for deliberation—be a representation-user and not just exhibit representation-isomorphic behavior—but it would need to *be* like us as well; otherwise, the world would not be significant-for them in the same way it is significant-for us and they would make the wrong decisions. The creation of such AI is not likely any time soon, but identifying the different gradations of potential systems will help us mark whether and how much progress is being made.

Categories of Systems

Non-Intentional Unilevel System

A non-intentional unilevel system has no action-initiating mechanism, and all of its processes are produced by mechanisms whose components sit at one constitutional level. The functional relation between these components is causal only.

Machine learning systems of all kinds are good examples of non-intentional unilevel systems. A Deep Belief Network (DBN), for example, is a deep learning system that is composed of stacks of single-layered learning networks. If the DBN's task is to parse images, for example, these component networks might be hierarchically organized so that the first one processes very small features of the image, the second one processes bundles of these, and so on until the macrofeatures of the image are identified. This structure is often described as multileveled, but this is misleading since there is no scalar difference between the levels and the relations between them are causal, not constitutive.

The relationship between sensory neural processes and perceptual neural ones is analogous: the former occur earlier and are conceptualized as lower than the latter, but they do not constitute different levels.

Non-Intentional Multilevel System

A non-intentional multilevel system has no action-initiating mechanism, but some of its processes are produced by mechanisms whose components sit at one or more constitutional levels. The functional relation between these components is constitutive and causal.

Weather systems, for example, tornadoes, and nonlinear chemical oscillators, for example, Belousov–Zhabotinsky reactions, are examples of these. They come together and dissipate according to the presence or absence of external control parameters, but they have no internal mechanism for maintaining system integrity. An ADAS is also an example of such a system since all cars are multilevel physical structures as described above; however, as already noted, its cognitive systems are not constitutively integrated across its levels.

Action-Intentional Unilevel System

An action-intentional unilevel system acts to maintain its own systemic integrity, and its processes are produced by mechanisms whose components sit at one constitutional level. The functional relation between these components is, therefore, causal only.

Single-celled organisms such as paramecia might be thought of as intentional unilevel systems.[8] Perhaps some of the virtual examples in evolutionary robotics are examples of these as well.

Action-Intentional Multilevel System

An action-intentional multilevel system has an action-initiating mechanism, and some of its processes are produced by mechanisms whose components sit at one or more constitutional level, but the stuff of its world is significant-for it in a largely hard-wired way. The functional relation between its mechanistic components is constitutive and causal.

It has been suggested (Edelman and Seth 2009; Griffin 1976; Grinde 2013) that having a centralized nervous system is a prerequisite for sentience and perhaps this is the right place to draw the line between simple action intentionality and directed intentionality. I imagine this line is a fuzzy one. Perhaps bacteria, fungi, and some invertebrates with very simple centralized nervous systems, such as mollusks and insects, might be examples of these systems.

Directed-Intentional Multilevel Systems

A directed-intentional multilevel system has an action-initiating mechanism and a perception mechanism with which it probes the environment for what is significant-for it. Its processes are produced by mechanisms whose components sit at one or more constitutional level.

Likely, many animals are examples of this kind of system; humans certainly are.

Conclusion

If autonomous AI systems are to integrate into the public space, we need to ensure they do so safely. On this externalist analysis of our most promising candidates, ADASs, the verdict is that they are not ready. Indeed, given that current top-down approaches do not concern themselves with the critical structural relations between a representation-intentional agent and its cognitive systems, we shouldn't hold our breath. Bottom-up approaches that recognize that action intentionality is the essential ground on which to build are more promising, but they are still a long way off on the cognitive front.

There is a larger question looming here: Why create autonomous AI systems at all? Tim Miller, who is spearheading a paradigm shift to what he calls "Evaluative AI,"[9] argues that current approaches will hinder us in the long run because they either undermine or misalign with human decision processes: "Research shows that people tend to either under-rely on tools, meaning they have no effect on decision making, or they over-rely on tools . . . , likely because providing recommendations makes them fixate on that recommendation. Both under- and over-reliance have negative consequences" (Miller 2023: 333–334). As the modern human is quite literally overwhelmed by information, improved data-processing technologies that will assist us in making better decisions are certainly needed. ADASs of the sort we already see on the roads have made enormous strides in this area by making driving for humans safer. If the motivation for creating autonomous AI is to help humanity, then, the best course of action would be to shift developmental focus away from creating systems that decide for humans and toward systems that provide support for (and thereby improve) human decision making.

If the motivation for creating autonomous AI is purely scientific, on the other hand, we are faced with some very large moral questions. If we succeed, what role will these new beings have in human society? Since they will be intentional systems with drives of their own, how will society accommodate their needs and desires? Our track record thus far in dealing with the intentional systems already on the planet is deplorable;[10] are we ready to add more to the mix? On balance, then, the rush to create autonomous AI systems is hasty, unwise, and likely to end badly.

12

Philosophical Implications

For philosophers who have thought deeply about intentionality, this account of Externalism will raise many questions. Here, I try to address some of these; there are many more.

Enactivist Representations?

Enactivists might balk at my appeal to their antirepresentational insights in the service of an account of representation-use. I hope that on closer inspection, however, that worry is allayed. On Externalism, a representation-intentional agent is not someone who has become imbued with inner representational states. Noë (2021) is right to complain that "It is a mistake . . . to concede, to the establishment, as it were, that these distinctive forms of thought are contentful and representationally based" (962). Agreed. Representational intentionality is the capacity to use external representations—words and such—to achieve things that would otherwise not have been possible. Sometimes we speak words silently to ourselves but, even then, we are using external tools of representation. Furthermore, although I've been focusing here on the strictly representational use of words, I do not intend to imply that this is the only or even the most usual way in which words are used. As I hope some of my scenarios have suggested, word-use is fundamentally something we do,[1] as Noë says, "a mode of activity and a medium for doing, thinking, finding, exploring" (2021). We use language to flirt, fight, soothe, cajole, dominate, support. Thus, I concur that language plays a much richer role in our lives than this account presents, but the intent here is to explain our representational capacity without assuming inner representational resources and the borders needed to be drawn somewhere. Further work using the externalist framework developed here, I hope, will include exploration of these non-representational uses of language.

Causation at Different Scalar Levels

Externalism analyzes relations within mechanisms according to function *and* structure, and one important consequence of this is that causal relations can hold at multiple levels of organization. Since reductionists[2] hold that multilevel causal analyses are misguided, a word needs to be said in defense of this approach.

Reductionists see high-level properties such as *being a mental state* as epiphenomenal consequences of low-level causal relations (Kim 1998: 60–7). Deliberations might *feel* like causes of behavior, but, for reductionists, all the causal work is done by "realizers," in our case, the nervous system. Indeed, the assumption that something like this reductive view must be right is the very source of our puzzle: how can reasons ever be *causes* of actions?

While reductionists are correct to insist that system states are constituted by subsystem ones, the bias toward treating the lowest constitutional level as *the level of causation* is tempting only when we try to account for the various phenomena in the world from a God's eye vantage point that privileges micro properties by fiat. From this "top-down" view on which the lowest organizational level has ontological primacy, system attributes are seen as *functional* properties that are realized by subsystem *physical* properties. But, as Glennan (2010) insightfully observes, this subsystem ontological bias is itself a consequence of a "tendency to think about the properties of objects as basic facts rather than mechanically explicable facts and a tendency to think about causal relationships as holding directly between properties rather than being mediated by particular mechanisms" (363). One of the guiding insights of Externalism is that properties are not basic facts; consequently, for complex explananda such as the behavior of organisms, the reductive picture is overly general.

When we adopt a mechanistic explanatory stance, in contrast, we take causal relevance to be determined by mechanistic *details.* What does this mechanism do? *How* does it do it? Glennan's example of a lock-and-key mechanism is a helpful paradigm case to keep in mind, and so I reproduce it in detail here.

> Cylinder locks have a mechanism involving a number of parts—most notably a cylinder in a casing with a slot for the key, and a set of pins that fall down into the cylinder and can be pushed up into the casing by the key. Key shape is causally relevant because it determines alignment of the pins, and proper alignment of the pins is what allows the cylinder to turn. Anyone can establish that key shape is a relevant property, simply by doing experiments with different shaped keys. But we do not understand why shape is relevant until we understand how the lock works. By understanding the parts and organization of the mechanism we are able to explain and predict why certain keys will work and others will not. (2010: 377–8)

Thus, whether a set of components sits at the bottom or top of a structural hierarchy within a given mechanism is not, on its own, an indicator of whether it plays a causal/functional role; rather, the structural details of the mechanism determine this. As we've seen, in complex multilevel systems, we can expect such causal relations at multiple scalar levels.

Doesn't Intentionality Presuppose a Capacity for Second-Order Thought?

Giving an account of intentionality is, many think (Kriegel 2009; Lycan 1996; Rosenthal 1998; Van Gulick 2004), just to give an account of a capacity for second-order thought. According to Externalism, this is right in one way and deeply mistaken in another. It is right to observe that a representational tool such as language gives one a capacity to talk and think about things. If "cat" stands for <*cat*>, then Fred talks about cats when he utters the word. But it does not follow from this that Fred now *has* a second-order capacity.

As Clark (2006b: 294) observes, even simple symbol tokens can convert difficult second-order tasks into manageable first-order ones. Language-naïve chimpanzees who are taught to associate first-order relations such as sameness and difference with symbol tokens—e.g., the "same" token for a banana + banana pairing and the "different" token for a banana + ball pairing—are subsequently capable of solving the hitherto opaque second-order task of applying those same relations to pairs of pairs—e.g., a "different" token for a pairing of banana + banana grouped with a pairing of banana + ball (Thompson, Oden, and Boysen 1997). These chimps have not thereby gained a new capacity, an inner understanding of some abstract relation; rather, with the aid of symbol tokens, the second-order problem is reduced to the first-order one of grouping symbol *tokens*—the banana + banana group is represented by the "same" token and the banana + ball group is represented by the "different" token, and the problem now is the simple one of seeing that "same" + "different" tokens are "different."

Language works similarly. Our brains aren't equipped for second-order relations, but with the help of words, we can solve problems that require them.

If Sally Intends to Eat Dessert Tonight, Isn't There Something About Sally Now That Represents This Intention?

What does this externalist account say of Sally, who plans to eat dessert after dinner tonight? For materialist internalists who are committed to realism about mental representations, content must be physically manifested in some way: if Sally plans to order dessert at the restaurant, there must be something about Sally now that represents this future intention. In other words, a successful reduction of mental content to lower-level physical processes, "vehicles of content," is a necessary condition of realism about the mental realm. But, as I argue in Chapter 6, this internalist program cannot succeed. Does this mean that mental representations are not real?

If what we mean by "real" is a version of "Fodor's industrial-strength realism" (Dennett 1991: 42), then, no, since on that view "beliefs and their kin would not be real unless the pattern dimly discernible from the perspective of folk psychology could also be discerned (more clearly, with less noise) as a pattern of structures within the brain" (Dennett 1991: 42). On the other hand, if what we mean by "real" mental representation includes something extended, that is, the result of a tool-dependent cognitive process, then, yes, mental representations are real: "The process that produces the data of folk psychology . . . is one in which the multidimensional complexities of the underlying processes are projected through linguistic behaviour, which creates an appearance of definiteness and precision, thanks to the discreteness of words" (Dennett 1991: 45).

Is there something now about Sally that represents her future intention to eat dessert? Yes and no. In having the intention, Sally uses words to express to herself that she will eat dessert tonight. This act of expression is itself constituted by activity at different levels of granularity: the low-level neural activity that is active when Sally is using the relevant words; the midlevel circulatory and/or digestive activity that is caused and constituted by neural associations with this activity; the high-level activity of repeatedly using words that Sally might engage in if her attention is particularly focused on her "intention"—"Yes, I'll have dessert tonight"—which in turn cause responses—episodic flashes of past dessert experiences, hunger, salivation—that serve to heighten her focus further. By the time Sally is at dinner, unless something more powerful has drawn her attention, the loop of activity will reoccur, perhaps less consciously if it has been deeply engrained, and Sally will fulfill this intention to eat dessert. At no point is there an inner intention that causes Sally to eat dessert; rather, the ongoing, extended, multi-layered activity together constitutes her intention to eat dessert. Externalism is thus realist about content in this extended sense, but eliminativist about "Real" content.

Much Ado about Non-Veridical Perception

Non-representationalist theories of perception have gained traction over the past decade and, taken as a family of "relational" views (Brewer 2017; Hurley 1998; Martin 2002; Noë2004; Noë 2010; O'Regan and Noë 2001; Travis 2013), offer fertile theoretical ground for Externalism. Noë's recent (2016, 2021) shift toward Sensorimotor Enactivism (SE), on which perception is an active and dynamic process according to which an organism interacts with its environment, is deeply synergistic with the externalist view I've been advancing here.

On SE, the manner in which perception is enacted is determined by a host of factors, including embodiment details—for example placement, size, maneuverability of the eyes, ears, nose, mouth—as well as modality-specific sensorimotor contingencies that guide and constrain an organism's perceptual activity. Perception is the ongoing exercising of these sensory modes of environmental engagement. There are no resultant, static percepts or awarenesses that a perceiver "has" on this view since "the outside world acts as an external memory that can be probed at will by the sensory apparatus" (O'Regan and Noë 2001: 946). Thus, there are no objects of perception, no (internal) objects of awareness either, only objects of engagement: S's awareness of X is an experience of X; S's perception of X is the activity of paying ongoing attention to X.

With respect to phenomenal aspect, what a sensory modality "is like" for an organism is largely a matter of the sensorimotor contingencies of the modality in question. For example, during vision, the sensory stimulation of retinal photoreceptors will shift and change in lawful ways as eyes rotate. The "flow of information" consequently will expand and contract as an organism moves forward and backward. Animals with eyes on the backs of their heads or on their knees will have different kinds of visual experiences than those with eyes on the front of their heads. The phenomenal aspect will vary across modalities as well. Unlike the tactile modality, visual stimuli are received only from the parts of objects that are facing an organism's visual sensory receptors. An organism skilled in vision and touch will develop expectations that accord with the idiosyncrasies of each of these modalities, for example, partial objects when visually searching but whole objects when touching. For different organisms, different expectations will develop.

If attentive awareness does not yield representations of the sensory data our receptors collect, we might wonder why we do not see gaps corresponding to our retinal blind spots, saccadic smears as our eyes literally jump around, and color fade-outs at the edges of our vision. On the representational view, these are all neatly explained by appeal to the reconstructive process that removes these by-products of our sensory apparatus. If we did not have such internal representations, wouldn't visual experience be of a jagged, discontinuous, Picassoesque world?

No. To be a perceiver is to be an organism skilled in interacting with its environment. When we are visually aware of an object, we see it without gaps because "the slightest flick of the eye, or of attention, renders it visible" (O'Regan and Noë 2001: 947). While perceiving, we interact with actual objects in the world, objects that are whole and continuously present. We move and adjust our bodies, our heads, our eyes so that we attend to different aspects of our locality to sense what is there. On this view, we never "have" a complete, uninterrupted visual experience; rather, we continually experience the world—interact with it—not at the subpersonal level, at which there are discrete, gappy moments of sensation, but at the system level, a vantage point from which the chaos of microlevel activity appears calm, smooth, and slow to change.

What then of non-veridical perception, when we look at a stump but mistakenly take it for a dog, or when we hallucinate and see/hear things that aren't there at all? One of the virtues of representationalism, so its proponents claim, is its simple explanation of such phenomena. Since perceptions have both representational content and phenomenal aspects, both veridical and non-veridical perceptions can have representational content "in common" and, sometimes, share phenomenal aspects as well: both my accurate perception of the mug on the desk and my mug hallucination are of the coffee mug and, to me, indistinguishable. The same representational mechanism underlies both, but in the veridical case, the representation is caused by the mug itself while, in the hallucination case, the representation is caused by something else, something internal to me. Representationalism thus offers a parsimonious account of both veridical and non-veridical perception.

Relationalists have responded to this challenge in a variety of ways, but the general strategy is disjunctivist: since hallucination does not involve interaction, while misperception does, different mechanisms must underlie them (Brewer 2006; Johnston 2004; Martin 2004). More recent predictive models have made interesting suggestions about what some of these different mechanisms might be (Clark 2016: 79–82); but, as this would take us far afield, I will leave the topic of hallucination there and describe briefly how misperception might be construed on an externalist view.

The first thing to note is that the very question presupposes some version of Representationalism: a non-veridical perception is one that does not match up accurately with what it represents. If we think of perception as an interactive process, on the other hand, something organisms do, then the question of accuracy does not even arise: can one act truly or falsely? We might act more or less skillfully, and this in turn might have consequences that could be cashed out in terms of truth and falsity, but then, as Brewer (2006) notes, it is the interpretation, a subject's judgment of its interactions, that is accurate or not: "Any errors in her world-view which result are products of the subject's responses to this experience" (169).

Suppose, for example, that a cat, on perceiving a distant stump in the forest, begins to hiss and exhibit a series of behaviors that we, third-person observers who study animal behavior, classify as <*defensive behavior*>. Why, we might wonder, is the cat responding in this way when no obvious threat is present? If we have observed repeated instances of cats exhibiting this behavior in the presence of dogs, we might offer up the hypothesis that the cat misperceives the stump to be a dog. This would link the cat's current behavior with past behavior in a way that makes its overall actions consistent.

On Externalism, however, this explanation does not entail that any of the cat's sensory processes (mis)represents <*dog*>. Although describing the behavior as a mistaken response to the current situation, as a misperception of the stump, is useful as an explanatory heuristic, we go awry when we look for some lower-level constituent of the cat's behavior about which we can say, "there, that's the <*dog*> representation that was mistakenly activated here." As I argue in Chapter 4, such d-representational glosses are accurate only as *very abstract* generalizations; they do not identify processes that play a representational role within the cat itself.

This is a break from tradition and common sense, to be sure. According to Representationalism, perception is the mechanism by which organisms receive information from their environments. The resulting percepts, on that view, are what are experienced. In looking at an apple, my perception is "as of" something small, round, and red—these are impressions in my mind. If Fred, walking through the forest at night, recoils upon seeing a distant stump and begins to change direction, it is Fred's perception of the stump "as of" a dog that causes Fred to change direction, that is, it is Fred's misrepresentation of the current situation that, at least partly, causes his behavior.

From the vantage point of the critique I have been developing here, however, such analyses confuse final and efficient causes. When we talk about Fred's perception of the stump "as of" a dog, we are generalizing over the entire bundle of spatiotemporally extended behavior beginning with his learning of a <*move away*> response to dogs and ending with his present <*move away*> response to a stump; this generalization does not exist *in* Fred. As Fred moves, the contours of the situation he is in extend spatiotemporally in different directions. Fred may have entered the forest in an anxious state, perhaps because he is uncomfortable walking through the forest at night by himself. Fred might have had a recent experience with a hostile dog, priming him for a response. Responses such as <*move away*>, connected as they are to past dog experiences, are attractors in the current situation, which is to say, they are response patterns that Fred will tend to enact. Thus primed, a distant, shadowy stump might indeed efficiently cause Fred to react with (what we describe as) a <*dog*> response. From the perspective of our final cause analysis—the explanatory reasons we use in our interpretation of the entire situation—we might say that something has gone wrong, that Fred has misperceived the stump as a dog. But from the perspective of occurrent Fred, it is the physical stump in Fred's environment that is the efficient cause of his <*move away*> response. In other words, it is perfectly consistent for Fred's perceiving the stump as a <*dog*> to explain/justify (final cause) his response without the perception of the stump as a <*dog*> thereby serving as an efficient cause of Fred's behavior. Indeed, much or most of our behavior is like this, that is, *not* caused by reasons.

This, at least, is the beginning of the externalist response to the "problem" of misperception. A representationalist might object that such accounts fail to explain *conscious* perceptual activity and the actions that seem to be guided by that activity: Fred, in misperceiving the stump "as of" a dog, becomes aware of the stump/dog and its potential threat, and this awareness seems to play a role in keeping Fred moving away from the stump. Moving away from the stump/dog is a goal for Fred, and his occurrent behavior of continuing to move in a trajectory away from the stump is most comprehensively viewed as implementing an algorithm for <*move away*>, in other words, as an efficient cause of his actions.

The externalist response is twofold: first, it does not follow from the fact that some of Fred's behavior is motivated by justifying reasons, that all of it must be so; second, the explanation of how some behavior is caused by reasons—in this case, by Fred's belief that there is a dog in the distance—requires a great deal of explanatory scaffolding. Externalism about representation-intentionality offers such an account.

Conclusions

Recall our opening puzzle: Sally is initially angry with Fred for ruining her garden but is mollified when he tells her that he did so accidentally and that he will repair the damage. On the standard view, when Fred says, "I promise to replant the garden," he communicates to Sally a picture *that he already has* of how the situation might unfold. Sally understands what Fred is saying because she translates/decodes his words into her own internal representations. On this view, the Problem of Intentionality is explaining what these internal representations are and how they communicate them to each other when they use language.

As I've been arguing, this approach puts the cart before the horse—words don't communicate representations; they *are* the representations we use to paint these past and future possibilities. Without the stabilizing and attention-anchoring support of language, thoughts are amorphous and ephemeral, not ground that can support deliberative reasoning. The Problem of Representational Intentionality, on this view, is explaining how we become skilled representation-*users*.

Externalism identifies both internal and external preconditions of this development: basic directed intentionality and a capacity for expectations^{sys} are necessary endogenous conditions of learning to use representations; having a rich linguistic community within which one develops this representation-using skill is a necessary exogenous condition of becoming a robust representation-user, someone capable of deliberation.

Over time, as an individual develops into a sophisticated word-user, a new capacity begins to emerge, one that is possible only with the aid of representation tools. Words *qua* objects draw our attention and trigger responses, but *qua* representations, they draw our attention to how they can be used. When we are perceptually engaged with words—speaking, reading, thinking—we are continually being drawn to the other ready-to-hand words in our environment. As we reach for them, we perpetuate our ongoing interaction with them: our attention is continually being drawn back to the words—they are tools that are *there*—and as we reach for more, we sustain the interaction. In the context of a deeply entrenched linguistic practice, words can function as a tool for focusing attention—verbal cues to "stay focused on the breath" make it easier to do so—but they can also be a large source of distraction since, with each utterance, new possibilities draw our attention hither and thither. Constant denizens of our internal landscape, word-objects bubble in and out of awareness. As our linguistic proficiency grows, words become increasingly ready-to-hand until, eventually, the interaction between self and words is a coupling that, for the tool-user, feels natural. We feel that we understand the words that we use.

The focus in mainstream cognitive science and analytic philosophy on representations is motivated by the important insight that a capacity to reason *about* things is critical to cognitive activity such as planning, problem-solving, and making decisions. Challenges from 5E cognitive science, on the other hand, are correct that, most fundamentally, humans (and non-human animals) interact with their world; they do not represent it. Externalism brings these two insights together. We are, at base, actors. However, when acting, we often use language, a tool for representation. With it, we can plan, solve complex problems, develop life-enhancing technologies, and learn from the experiences of others. This skill helps us to survive but also comes with responsibility: when our actions are thus deliberatively grounded, we are responsible for their consequences.

Notes

Introduction

1 Some of the material in this book appears, by permission, in (Salay 2019) and (Salay 2021), two papers that represent the germ of the ideas that I develop here.
2 More precisely, it is the Problem of Representational Intentionality. In Chapter 5, I identify two more basic kinds of intentionality: Action and Directed Intentionality.
3 This sense of the word, to plan to do something, is not to be confused with the technical notion of intentionality that is the topic of this book, though of course there is a close relation between the two senses that will become clearer.
4 As the Problem of Intentionality is generally treated as the Problem of Representational Intentionality by analytic philosophers, it is this group that has offered up theories to account for it. Some notable ones are Dretske (1981, 1995); Fodor (1987); Millikan (1984); Neander (2017); and Shea (2018).
5 This idea is not new: see Clark (2006a, 2006b, 2015); Donald (1991); Jeffares (2010); Logan (2007); Menary (2007); Rowlands (2006); Sterelny (2010); Sutton et al. (2010); and Wheeler (2004, 2005, 2011, 2013, 2015, 2019).

Chapter 1

1 Internalists do. More on this in Chapter 6.
2 A rigorous conceptual analysis approach to philosophical inquiry.
3 A rigorous observational method for studying consciousness from the subjective perspective.
4 Namely, embodied, enactive, extended, embedded, and ecological.
5 Some aspects of the world will be "unready-to-hand," of course, but these too are the result of a more basic being-in-the-world for Heidegger.
6 Noteworthy contributions to an enactivist approach to cognition include Chemero (2009); De Jaegher and Di Paolo (2007); Di Paolo (2009); Di Paolo, Cuffari, and De Jaegher (2009); Gallagher (2017); Hutto and Myin (2012; 2017); Noë (2009, 2021); and Stewart, Gapenne and Di Paolo (2010).

Chapter 2

1 Hilary Putnam's (1967) is one of the first explicit formulations of this.
2 Esther Thelen and Linda Smith's seminal work on developmental dynamics (1994), for example, is now required reading in Developmental Psychology and Varela, Thompson, and Rosch's, now classic, *The Embodied Mind* (1992) is one of the most cited books in the field.
3 In Chapter 11, I give an externalist analysis of current AI systems, and there we see the full consequences of this problem.
4 Important works in this area include Bechtel (2011); Chemero and Silberstein (2008); Craver (2001, 2006, 2007); Cummins (1975); Fagan (2012); Glennan (2010); Levy and Bechtel (2013); and Weiskopf (2011).

Chapter 3

1 See, for example, Bechtel and Abrahamsen (2010); Beer (1995, 2000); Bishop (2008); Bressler and Kelso (2001); Chemero (2009); Chemero and Silberstein (2008); Collier and Hooker (1999); Cosmelli, Lachaux, and Thompson (2007); Di Paolo (2005); Gánti (2003); Keijzer (2002); Keijzer, Van Duijn, and Lyon (2013); Kelso (1995); Kelso and Engstrom (2006); Port and Van Gelder (1995); Silberstein (2002); and Thelen et al. (2001).
2 Ezequiel Di Paolo (2005) develops this idea in conjunction with the concept of adaptivity, and this, as we shall see in Chapter 11, is a key design principle of Enactivist AI.
3 There is a synergy between Heidegger's notion of being-in-the-world, transcendence, and action intentionality that needs further consideration: "Intentionality is founded in Dasein's transcendence and is possible solely for this reason—transcendence cannot conversely be explained in terms of intentionality" (Heidegger 1982: 162).
4 See this documentary for an example of this confusion in action: https://www.academy.ca/2017/my-brain-made-me-do-it/.
5 These ideas are intended to complement Lynne Rudder Baker's notion of constitution: "The fundamental idea of constitution is this: when a thing of one primary kind is in certain circumstances, a thing of another primary kind—a new thing, with new causal powers—comes to exist" (Baker 2007: 32).

Chapter 4

1 Significant contributions that have deepened the collective understanding of representation include (Chemero 2009; Egan 2014; Haugeland 2000; Markman and Dietrich 2000; Ramsey 2003, 2007).

2 Although I will not specifically address the question of which model of perception best complements Externalism, I note here that an account that focuses on the interactive role of perception, as the enactivist sensorimotor view does, will be a better fit than the standard information processing one, for reasons that should become clear.

3 As will become clear in Chapter 6, two different senses of representation are appealed to in these accounts. This, I will argue, is a mistake.

4 The situation is not only a representational one of course. There are direct, physical aspects of the interaction: Sally's mirror neurons are triggered by Fred's facial and bodily expressions, for example, and she might begin to feel his sympathy. But we are only interested here in the part of the exchange that is representational.

5 I give this explanation in Chapter 7.

6 In terms of Ramsey's (2007) analysis, both receptor-representation—X represents Y by way of being a reliable indicator of Y—and s-representation—X represents Y in virtue of being isomorphic with Y—are both instances of a d-rep relation, and io-representation—X represents Y when S interprets X as Y—is an instance of a t-rep relation.

7 Henceforth, I will use the phrase "significant to" as a superclass of "significant-in" and "significant-for" when appropriate to improve narrative flow.

8 As I've already noted, if we do adopt this representational stance to such correlations, we must be careful to qualify them as d-representations, dyadic relations between internal microprocesses and macro objects, so that we don't confuse them with the t-representations we are investigating. There may indeed be an important relationship between d-rep relations of this sort and the t-rep relations we are trying to explain, but there may not. We must keep the two separate until we have a clear account of the latter.

9 An oversimplification that could be helpful here is the concept of homeostasis: an autopoietic system acts to achieve a balance between needs and resources.

10 In Chapter 8, I will argue that it is our intentional use of representations that grounds this attunement to objectivist objects such as <*Toronto*> and <*traffic*> and keeps us focused on such planning tasks.

Chapter 5

1 See Chapter 3 for more discussion on this.

2 See Chapter 11 for more discussion on this.

3 I want to note here that I am very sensitive to the wise cautioning of Lyon and Keijzer (2022) against an anthropogenic approach to cognitive science in which we determine the line between cognitive and non-cognitive according to some human baseline. I concur absolutely that we must guard against this. The thought driving the distinctions I make here is the observation that increasingly sophisticated sensory-motor systems are more complex in a way that seems relevant to understanding how mammalian behavior is more cognitive than that of insects. In contrast, increasingly sophisticated digestive systems, for example, do not seem relevant. Flexibility in directedness, then, might be a more biogenically inspired "mark of the cognitive" than representation-use is.

4 Some notable collections include the following: Gallagher and Zahavi (2020); Schmicking and Gallagher (2010); and Zahavi (2012).

Chapter 6

1 Not to be confused with "Constructive Perception Theory," the view that top-down context-fixing processes and bottom-up sensory ones merge together to yield a contextually guided percept. I am only interested here in the bottom-up part of the perception process as standardly conceived.

2 As I am using Shea's account in the service of fleshing out a relevant kind of reductive internalism, I should note that he explicitly describes his view as a *non-reductive*, physicalist account of content, because it purports to establish supervenience rather than identity relations between content and vehicle (41). Furthermore, it is partly externalist in the sense that it espouses a wide, relational view of content (39). This is all for the good. Nevertheless, it treats content specification as a matter of mapping outer patterns to inner physical mechanisms that function as vehicles of content, and it is this identification that I am arguing is confused.

3 As I argue in (Salay 2021), however, (Shea 2018) is not ultimately successful in its larger aim of showing that cognitive representations are the ground of intentionality.

Chapter 7

1 The Hard Problem too is simplified by this narrowing of scope.

2 Learning in affectively charged situations can yield one-shot learning, which means that very few experiences of the conjunction between fire and smoke are needed before the association is learned.

3 As Grice (1957) famously pointed out, smoke is a natural sign of fire. Indeed, it is, but as this account shows, not just any system can *exploit* this significatory relationship.

4 Although (Gallagher 2023) sheds some doubt on whether PP models will complement the wider, enactivist, organism view, perhaps an enactivist model of the brain as suggested by (Gallagher et al. 2013) is needed as well.

5 Though, for all the reasons I gave in Chapter 4, treating neural correlations as representations of any sort can lead to a deeply skewed view of system behavior.

6 We could say that N_1 expects N, but this wouldn't be very helpful in an account of S's representation-using behavior.

7 In philosophy, the roots of this distinction extend much farther back, at least to Gilbert Ryle's *The Concept of Mind* (1949), in which he gave an analysis of the difference between explicit propositional knowledge—e.g., knowing that it is now 8:00 am—and implicit embodied knowledge—e.g., knowing how to tie one's shoes.

8 There is debate about whether or not insects are sentient (Ray 2018; Hill 2016; Klein and Barron 2016). For the purpose of this example, I assume they are not, but nothing in this account hinges on this being true: if crickets are sentient and capable of learning responses to t-rep relations, then they are capable of direct representation-use.

9 Specifically, this means that "synaptic connection from 'CS' neurons to 'OA/DA' [octopaminergic/dopaminergic] neurons representing unconditioned stimulus (US), which is termed CS-US or S-S connection [19–23], is strengthened by coincident activation of 'CS' neurons and 'OA/DA' neurons during conditioning" (Mizunami et al. 2009: 7).

10 For discussions and investigations into whether and to what degree vervet calls constitute a language, see (Macedonia and Evans 1993; Price et al. 2015; Seyfarth, Cheney, and Marler 1980; Struhsaker 1967).

Chapter 8

1 An externalist would say rather "to *do* things with words."

2 John Haugeland's slogan that "If you take care of the syntax, the semantics will take care of itself" (2012: 51) was the rallying cry of GOFAI.

3 The developmental trajectory of language learning for children begins with visual attentiveness—noticing the connection between words and their objects (Tomasello 2005)—and sound mimicry, which supports the critical step of learning a language's prosodic flow (Esteve-Gibert and Prieto 2018).

4 Although the analysis of word meaning in analytic philosophy of language suffers because the approach is overly focused on *functional* relations and ignores the messy but important details of word *use*, David Lewis's (1969: 35–42) formalization of solutions to coordination problems is noteworthy and helpful (except, of course, that there is no need to appeal, as he does, to internal representational states!).

5 See Esteve-Gibert and Prieto 2018) for a compilation of important recent work in the field of prosody development, a prelinguistic developmental period during which language learners become tuned to the rhythm and cadence of a language, a critical foundation for identifying word beginnings and endings.

6 Reading about or being taught the proper usage of a concept is itself a learning situation.

7 That self-talk is used in this way has been extensively studied and, in a review of the developmental literature, Zelazo (2015) concludes that

> Contemporary research generally supports the seminal ideas of Vygotsky (1962) and Luria (1959, 1961) concerning the importance of verbal processes in the exercise and development of self-regulation, finding, for example, that with age children increasingly use verbalization strategically to maintain task information in mind (Karbach and Kray 2007), and that blocking the use of inner speech disrupts cognitive control in children and adults. (Emerson and Miyake 2003; Kray, Eber, and Karbach 2008) (61).

8 I draw, here, on Savage-Rumbaugh's (1990) descriptions of Kanzi.

9 Lexigrams are visual symbols for words that are used in language instruction with nonverbal animals such as great apes.

10 When we consider the results of human attempts to teach language to nonhuman animals, we should keep in mind that the words being taught are humancentric ones, concepts that pick out what is significant-for humans in the context of our physical and social environments, rather than what is significant-for them, for example, terms for different kinds of dominance behavior. A further factor, of course, will be how easy or difficult production in a given sign medium is for the subject. Finally, but certainly not comprehensively, consideration must be given to the learning conditions—if learning takes place in an unnatural, captive

setting, the learner will be experiencing anxiety and stress, factors that we know inhibit learning. For a review of the way that anxiety inhibits language learning specifically, see (Horwitz 2001).

11 Again, in some scientific contexts, it is useful to think about Fred's perceptual activity as constructing a representation of the flower. But these are d-representations only, and not the t-representations that Fred uses when he thinks and speaks about the flower.

12 Willard Sellars (1956) and Jay Garfield (2016) explain this nuanced distinction beautifully.

Chapter 9

1 This is a burgeoning field. Here is a sampling of some recent work: Boldt and Mortensen (2022); Galke, Ram, and Raviv (2022); Lemon (2022); Ossenkopf, Luck, and Mathewson (2022); Ueda et al. (2022).

Chapter 10

1 In Chapter 12, I give a justification for treating causation as a nonreducible, intralevel relation.

2 As I discuss in Chapter 8, however, a capacity for becoming aware of expectations^{sys} *is* a condition of sophisticated representation-use.

3 I was introduced to these studies by this wonderful review paper (Michael, Garry, and Kirsch 2012). I recommend it to those interested in learning more about the various ways in which our expectations^{sys} direct our actions at multiple scalar levels.

4 See, for example, Atlas and Wager (2012); Montgomery and Kirsch (1999); Mora, Nestoriuc, and Rief (2011); Tracey (2010); Zubieta and Stohler (2009).

5 See, for example, Benedetti et al. (2003); Benedetti (2013); Bingel et al. (2011); Colagiuri et al. (2015); Colloca and Benedetti (2005, 2009); de la Fuente-Fernández et al. (2001); Kaptchuk et al. (2010); Kaptchuk and Miller (2015); Kirsch, (1997), Price, Finniss, and Benedetti (2008), Wechsler et al. (2011).

6 See, for example, Alabas et al. (2012); Aronson and Inzlicht (2004); Beilock, Rydell, and McConnell (2007); Barber and Mather (2013); Defrin, Eli, and Pud (2011); Derks et al. (2011); Forbes et al. (2015, 2008); Forbes and Leitner (2014);

Fowler et al. (2022); Friedrich et al. (2015); Huber et al. (2015); John-Henderson, Rheinschmidt, and Mendoza-Denton (2015); Krendl et al. (2015); Lamont, Swift, and Abrams (2015); Laurin (2013); Matheson and Cole (2004); Racine et al. (2012); Schwarz (2015); Schmader, Johns, and Forbes (2008); Stoet and Geary (2012); Wraga et al. (2006); Sanford et al. (2002); Blascovich et al. (2001); Robinson et al. (2001, 2003); Rosenthal and Jacobson (1966); Stone et al. (1999); and Steele and Aronson (1995).

7 See, for example, Altmann and Kamide (1999); Aviezer et al. (2008, 2011); Barrett and Kensinger (2010); Bruner and Minturn (1955); DeLong et al. (2005); Diekhof et al. (2011); Kim et al. (2004); Kok et al. (2013); Long and Toppino (2004); Ofan et al. (2014); Riccio, Cole, and Balcetis (2013); Righart and Gelder (2008); Schwarz (2015); Sedivy et al. (1999); Wieser and Brosch (2012); Wieser et al. (2014).

8 The authors note that although this is already a well-established phenomenon in the field—"Mental imagery or 'seeing with the mind's eye' holds the power to build up vivid internal 'as if' representations, which enable a mental simulation of actual visceral and emotional responses (Bensafi et al. 2007; Dadds et al. 1997; Kim et al. 2007; Kosslyn et al. 1996; Vrana and Lang 1990)" (Diekhof et al. 2011: 1704)—"the actual role of anticipatory neural processing in the generation of illusory percepts and in the reduction of negative affect has been widely neglected" (Diekhof et al. 2011: 1704).

Chapter 11

1 These were spearheaded by Rodney Brooks (1986) under the heading of "subsumptive architectures."

2 Notable contributions to this field include: Beer (1995, 1997, 2000); Di Paolo (2003); Froese and Ziemke (2009); Macinnes and Di Paolo (2006); Pfeifer, Iida, and Bongard (2005); Wood and Di Paolo (2007); and Ziemke (2005).

3 This is called Level 5 ADAS in the SAE Standard J3016_202104, the most recent version of the *Taxonomy and Definitions for Terms Related to Driving Automation Systems for On-Road Motor Vehicles* put out by SAE International.

4 Unsupervised training is also possible, but as the method by which DNNs are trained typically occurs offline, this isn't relevant to an analysis of ADAS systems.

5 Tim van Gelder (1995) makes this important point in the context of the Watt Governor, a mechanism for maintaining a loom at a constant speed: "All behavior in the centrifugal governor happens in the very same real-time frame as change in the speed of the flywheel" (355).

6 They are not significant-in R since S is only causally, not constitutively, related to R. Indeed, they are only significant-in the DNN that is trained to classify it.
7 See (Di Paolo 2005) for an explication of this idea.
8 I offer this example reluctantly since it is a paramecium's subcellular processes that together implement its action intentionality.
9 A related shift toward "Explainable AI"—see, for example (Hu 2023; Zhou and Hu 2023)—replaces "black-box" models with those that human users can probe so that they can judge how/why an AI system's output is arrived at.
10 According to a 2019 UN Report, to date at least 680 vertebrate species have become extinct due to human activity and "1 million animal and plant species are now threatened with extinction" (https://www.un.org/sustainabledevelopment/blog/2019/05/nature-decline-unprecedented-report).

Chapter 12

1 (Di Paolo, Cuffari, and De Jaegher 2019) is a detailed analysis of this idea.
2 Most famously, Jaegwon Kim (1998).

References

Alabas, O.A., Tashani, O.A., Tabasam, G., and Johnson, M.I. (2012), Gender role affects experimental pain responses: A systematic review with meta-analysis, *European Journal of Pain*, 16 (9): 1211–23.

Alon, U. (2007), Network motifs: Theory and experimental approaches, *Nature Reviews Genetics*, 8 (6): 450–61.

Altmann, G.T. and Kamide, Y. (1999), Incremental interpretation at verbs: Restricting the domain of subsequent reference, *Cognition*, 73 (3): 247–64.

Ardila, A., Bertolucci, P.H., Braga, L.W., Castro-Caldas, A., Judd, T., Kosmidis, M.H., Matute, E., Nitrini, R., Ostrosky-Solis, F., and Rosselli, M. (2010), Illiteracy: The neuropsychology of cognition without reading, *Archives of Clinical Neuropsychology*, 25 (8): 689–712.

Aronson, J. and Inzlicht, M. (2004), The ups and downs of attributional ambiguity: Stereotype vulnerability and the academic self-knowledge of African American college students, *Psychological Science*, 15 (12): 829–36.

Assefi, S.L. and Garry, M. (2003), Absolut® memory distortions: Alcohol placebos influence the misinformation effect, *Psychological Science*, 14 (1): 77–80.

Atlas, L.Y. and Wager, T.D. (2012), How expectations shape pain, *Neuroscience Letters*, 520 (2): 140–148.

Automated Driving Vehicle Approved as Automated Operation Level 4 System for the First Time in Japan. (2023), Joint Press Release with the Ministry of Land, Infrastructure, Transport and Tourism. (Retrieved May 20, 2024).

Aviezer, H., Bentin, S., Dudarev, V., and Hassin, R.R. (2011), The automaticity of emotional face-context integration, *Emotion*, 11 (6): 1406–14.

Aviezer, H., Hassin, R.R., Ryan, J., Grady, C., Susskind, J., Anderson, A., Moscovitch, M., and Bentin, S. (2008), Angry, disgusted, or afraid? Studies on the malleability of emotion perception, *Psychological Science*, 19 (7): 724–32.

Baker, L.R. (2007), *The metaphysics of everyday life*. Cambridge: Cambridge University Press.

Barber, S.J. and Mather, M. (2013), Stereotype threat can both enhance and impair older adults memory, *Psychological Science*, 24 (12): 2522–9.

Barrett, L.F. and Kensinger, E.A. (2010), Context is routinely encoded during emotion perception, *Psychological Science*, 21 (4): 595–9.

Bechtel, W. (2011), Mechanism and biological explanation, *Philosophy of Science,* 78 (4): 533–57.

Bechtel, W. (2013), Addressing the vitalists challenge to mechanistic science: Dynamic mechanistic explanation, *Vitalism and the Scientific Image in Post-Enlightenment Life Science*, 1800–2010: 345–70.

Bechtel, W. and Abrahamsen, A. (2010), Dynamic mechanistic explanation: Computational modeling of circadian rhythms as an exemplar for cognitive science, *Studies in History and Philosophy of Science Part A,* 41 (3): 321–33.

Beecher, M.D. (2021), Why are no animal communication systems simple languages?, *Frontiers in Psychology*, 12: 701.

Beer, R.D. (1995), A dynamical systems perspective on agent-environment interaction, *Artificial Intelligence*, 72 (1–2): 173–215.

Beer, R.D. (1997), The dynamics of adaptive behavior: A research program. *Robotics and Autonomous Systems*, 20 (2–4): 257–89.

Beer, R.D. (2000), Dynamical approaches to cognitive science, *Trends in Cognitive Sciences*, 4 (3): 91–9.

Beilock, S.L., Rydell, R.J., and McConnell, A.R. (2007), Stereotype threat and working memory: Mechanisms, alleviation, and spillover, *Journal of Experimental Psychology: General*, 136 (2): 256–76.

Benedetti, F. (2013), Placebo and the new physiology of the doctor-patient relationship, *Physiological Reviews*, 93 (3): 1207–46.

Benedetti, F., Pollo, A., Lopiano, L., Lanotte, M., Vighetti, S., and Rainero, I. (2003), Conscious expectation and unconscious conditioning in analgesic, motor, and hormonal placebo/nocebo responses, *Journal of Neuroscience*, 23 (10): 4315–23.

Bergman, T.J., Beehner, J.C., Painter, M.C., and Gustison, M.L. (2019), The speech-like properties of nonhuman primate vocalizations, *Animal Behaviour*, 151: 229–37.

Bermpohl, F., Pascual-Leone, A., Amedi, A., Merabet, L.B., Fregni, F., Gaab, N., Alsop, D., Schlaug, G., and Northoff, G. (2006), Dissociable networks for the expectancy and perception of emotional stimuli in the human brain, *Neuroimage*, 30 (2): 588–600.

Bialystok, E. (2001), *Bilingualism in development: Language, literacy, and cognition*. Cambridge: Cambridge University Press.

Bingel, U., Wanigasekera, V., Wiech, K., Ni Mhuircheartaigh, R., Lee, M.C., Ploner, M., and Tracey, I. (2011), The effect of treatment expectation on drug efficacy: Imaging the analgesic benefit of the opioid remifentanil, *Science Translational Medicine*, 3 (70), 70ra14.

Bishop, R.C. (2008), Downward causation in fluid convection, *Synthese*, 160: 229–48.

Blascovich, J., Mendes, W.B., Hunter, S.B., Lickel, B., and Kowai-Bell, N. (2001), Perceiver threat in social interactions with stigmatized others, *Journal of Personality and Social Psychology*, 80 (2): 253.

Boldt, B. and Mortensen, D. (2022), Modeling emergent lexicon formation with a self-reinforcing stochastic process. In *Emergent Communication Workshop at ICLR 2022*. April 2022.

Boly, M., Balteau, E., Schnakers, C., Degueldre, C., Moonen, G., Luxen, A., Phillips, C., Peigneux, P., Maquet, P., and Laureys, S. (2007), Baseline brain activity fluctuations predict somatosensory perception in humans, *Proceedings of the National Academy of Sciences*, 104 (29): 12187–92.

Bowers, J.S., Malhotra, G., Dujmović, M., Montero, M.L., Tsvetkov, C., Biscione, V., Puebla, G., Adolfi, F., Hummel, J.E., Heaton, R.F., and Evans, B.D. (2023), Deep problems with neural network models of human vision, *Behavioral and Brain Sciences*, 46: e385.

Bramao, I., Mendon, A., Faísca, L., Ingvar, M., Petersson, K.M., and Reis, A. (2007), The impact of reading and writing skills on a visuo-motor integration task: A comparison between illiterate and literate subjects, *Journal of the International Neuropsychological Society*, 13 (2): 359–64.

Brentano, F. (1874), *Psychology from an empirical standpoint*. London: Routledge and Kegan Paul.

Bressler, S.L. and Kelso, J.S. (2001), Cortical coordination dynamics and cognition, *Trends in Cognitive Sciences*, 5 (1): 26–36.

Brewer B. (2006), Perception and content, *European Journal of Philosophy*, 14 (2): 165–81.

Brewer, B. (2017), The object view of perception, *Topoi*, 36: 215–27.

Brooks, R.A. (1986), A robust layered control system for a mobile robot, *IEEE Journal of Robotics and Automation*, RA-2: 14–23.

Brooks, R.A. (1991A), Intelligence without representation, *Artificial Intelligence*, 47 (1–3): 139–60.

Brooks, R.A. (1991B), Intelligence without reason, *A.I. Memo No. 1293 MIT-AI Laboratory*: 1–27.

Bruner, J.S. and Minturn, A.L. (1955), Perceptual identification and perceptual organization, *The Journal of General Psychology*, 53 (1): 21–8.

Carlsson, K., Petrovic, P., Skare, S., Petersson, K.M., and Ingvar, M. (2000), Tickling expectations: Neural processing in anticipation of a sensory stimulus, *Journal of Cognitive Neuroscience*, 12 (4): 691–703.

Carr, J.W., Smith, K., Cornish, H., and Kirby, S. (2017), The cultural evolution of structured languages in an open-ended, continuous world, *Cognitive Science*, 41 (4): 892–923.

Castro-Caldas, A., Miranda, P.C., Carmo, I., Reis, A., Leote, F., Ribeiro, C., and Ducla-Soares, E. (1999), Influence of learning to read and write on the morphology of the corpus callosum, *European Journal of Neurology*, 6 (1): 23–8.

Cen, A., Yosinski, J., and Clune, J. (2015), Deep neural networks are easily fooled: High confidence predictions for unrecognizable images. *Proceedings of the IEEE Conference on Computer Vision and Pattern Recognition*: 427–36.

Chalmers, D.J. (1995), Facing up to the problem of consciousness, *Journal of Consciousness Studies*, 2 (3): 200–19.

Chemero, A. (2009), *Radical embodied cognitive science*. Cambridge: MIT Press.

Chemero, A. and Silberstein, M. (2008), After the philosophy of mind: Replacing scholasticism with science, *Philosophy of Science*, 75 (1): 1–27.

Chomsky, N. (1968), *Language and mind*. New York: Harcourt, Brace and World.

Christiansen, M. and Chater, N. (2008), Language as shaped by the brain, *Behavioral and Brain Sciences*, 31 (5): 489–509.

Christiansen, M. H. and Chater, N. (2015), The language faculty that wasnt: A usage-based account of natural language recursion, *Frontiers in Psychology*, 6: 1182.

Churchland, P. (1985), Reduction, qualia, and the direct introspection of brain states, *The Journal of Philosophy*, 82 (1): 8–28.

Cimatti, F. and Vallortigara, G. (2015), So little brain, so much mind. Intelligence and behaviour in nonhuman animals, *Reti, Saperi, Linguaggi*, 2 (1): 5–22.

Clark, A. (1997), *Being there: Putting brain, body, and world together again.* Cambridge: MIT Press.

Clark, A. (2002), Skills, spills and the nature of mindful action, *Phenomenology and the Cognitive Sciences*, 1: 385–7.

Clark, A. (2006a), Language, embodiment, and the cognitive niche, *Trends in Cognitive Science*, 10 (8): 370–4.

Clark, A. (2006b), Material symbols, *Philosophical Psychology*, 19 (3): 291–307.

Clark, A. (2015), Predicting peace: The end of the representation wars, in T. Metzinger and J.M. Windt (eds.), *Open MIND*: *7*(R), Frankfurt: MIND Group.

Clark, A. (2016), *Surfing uncertainty: Prediction, action, and the embodied mind.* Oxford: Oxford University Press.

Clark, A. and Chalmers, D. (1998), The extended mind, *Analysis*, 58: 7–19.

Clark, A. and Toribio, J. (1994), Doing without representing? *Synthese*, 101: 401–31.

Clifasefi, S.L., Garry, M., Harper, D.N., Sharman, S.J., and Sutherland, R. (2007), Psychotropic placebos create resistance to the misinformation effect, *Psychonomic Bulletin & Review*, 14: 112–17.

Colagiuri, B., Livesey, E.J., and Harris, J.A. (2011), Can expectancies produce placebo effects for implicit learning?, *Psychonomic Bulletin & Review*, 18: 399–405.

Colagiuri, B., Schenk, L.A., Kessler, M.D., Dorsey, S.G., and Colloca, L. (2015), The placebo effect: From concepts to genes, *Neuroscience*, 307: 171–90.

Collier, J.D. and Hooker, C.A. (1999), Complexly organised dynamical systems, *Open Systems & Information Dynamics*, 6 (3): 241–302.

Colloca, L. and Benedetti, F. (2005), Placebos and painkillers: Is mind as real as matter?, *Nature Reviews Neuroscience*, 6 (7): 545–52.

Colloca, L. and Benedetti, F. (2009), Placebo analgesia induced by social observational learning, *PAIN*, 144 (1–2): 28–34.

Cosmelli, D., Lachaux, J.P., and Thompson, E. (2007), Neurodynamics of consciousness, *The Cambridge Handbook of Consciousness*, 2: 229–39.

Craver, C.F. (2001), Role functions, mechanisms, and hierarchy, *Philosophy of Science*, 68 (1): 53–74.

Craver, C.F. (2006), When mechanistic models explain, *Synthese*, 153 (3): 355–76.

Craver, C.F. (2007), *Explaining the brain: Mechanisms and the mosaic unity of neuroscience*. Oxford: Oxford University Press.

Craver, C. and Bechtel, W. (2007), Top–down causation without top–down causes, *Biology and Philosophy*, 22: 547–63.

Croizet, J.C., Després, G., Gauzins, M.E., Huguet, P., Leyens, J.P., and Méot, A. (2004), Stereotype threat undermines intellectual performance by triggering a disruptive mental load, *Personality and Social Psychology Bulletin*, 30(6): 721–31.

Culbert, B.M. and Balshine, S. (2019), Visual threat signals influence social interactions in a cooperatively breeding fish, *Animal Behaviour*, 151: 177–84.

Cummins, R. (1975), Functional analysis, *The Journal of Philosophy*, 72 (20): 741–65.

Cuthbertson, A. (2024), Two driverless cars crash into the same truck, prompting first Waymo recall. *The Independent*. https://www.independent.co.uk/tech/driverless-car-waymo-recall-b2496141.html. (Retrieved May 20, 2024).

Davidson, D. (1975), Thought and talk, in S. Guttenplan (ed.), *Mind and language*, 7–23, Oxford: Oxford University Press.

De la Fuente-Fernández, R., Ruth, T.J., Sossi, V., Schulzer, M., Calne, D.B., and Stoessl, A.J. (2001), Expectation and dopamine release: Mechanism of the placebo effect in Parkinsons disease, *Science*, 293 (5532): 1164–6.

Deacon, T.W., 1998. *The symbolic species: The co-evolution of language and the brain. WW Norton & Company.*

Deacon, T.W. (2011), *Incomplete nature: How mind emerged from matter*. New York: WW Norton & Company.

Defrin, R., Eli, I., and Pud, D. (2011), Interactions among sex, ethnicity, religion, and gender role expectations of pain, *Gender Medicine*, 8 (3): 172–83.

De Jaegher, H. and Di Paolo, E.A. (2007), Participatory sense-making: An enactive approach to social cognition. *Phenomenology and the Cognitive Sciences*, 6 (4), 485–507.

DeLong, K.A., Urbach, T.P., and Kutas, M. (2005), Probabilistic word pre-activation during language comprehension inferred from electrical brain activity, *Nature Neuroscience*, 8 (8): 1117–21.

Dennett, D.C. (1991), Real patterns, *The Journal of Philosophy*, 88 (1): 27–51.

Dennett, D. (2001), Consciousness: How much is that in real money?, in R. Gregory (ed.), *Oxford companion to the mind*, Oxford: Oxford University Press.

Derks, B., Van Laar, C., Ellemers, N., and De Groot, K. (2011), Gender-bias primes elicit queen-bee responses among senior policewomen, *Psychological Science*, 22 (10): 1243–49.

Di Paolo, E.A. (2003), Organismically-inspired robotics: Homeostatic adaptation and teleology beyond the closed sensorimotor loop, in *Dynamical systems approach to embodiment and sociality*, 19–42, Adelaide: Advanced Knowledge International.

Di Paolo, E.A. (2005), Autopoiesis, adaptivity, teleology, agency, *Phenomenology and the Cognitive Sciences*, 4 (4): 429–52.

Di Paolo, E.A. (2009), Extended life, *Topoi*, 28 (1): 9–21.

Di Paolo, E. A., Cuffari, E. C., and De Jaegher, H. (2019), *Linguistic bodies: The continuity between life and language*, Cambridge: The MIT Press.

Di Paolo, E. A., Rohde, M., and De Jaegher, H. (2009), Horizons for the enactive mind: Values, social interaction, and play, in J. Stewart, O. Gapenne and E.A. Di Paolo (eds.), *Enaction: Towards a new paradigm for cognitive science*, Cambridge: The MIT Press, 33-87.

Diekhof, E.K., Kipshagen, H.E., Falkai, P., Dechent, P., Baudewig, J., and Gruber, O. (2011), The power of imagination—How anticipatory mental imagery alters perceptual processing of fearful facial expressions, *NeuroImage*, 54 (2): 1703–14.

Donald, M., 1993. *Origins of the modern mind: Three stages in the evolution of culture and cognition. Harvard University Press.*

Dretske, F. (1981), *Knowledge and the flow of information*, Cambridge: The MIT Press.

Dretske, F. (1995), *Naturalizing the Mind*, Cambridge: The MIT Press.

Dreyfus, H. (1976), What computers can't do, *British Journal for the Philosophy of Science*, 27 (2): 177–85.

Dreyfus, H. (2007), Why Heideggerian AI failed and how fixing it would require making it more Heideggerian, *Artificial Intelligence*, 171: 1137–60.

Edelman, D.B. and Seth, A.K. (2009), Animal consciousness: A synthetic approach, *Trends in Neurosciences*, 32 (9): 476–84.

Egan, F. (2014), How to think about mental content, *Philosophical Studies*, 170: 115–35.

Esteve-Gibert, N. and Prieto, P. (2018), *The development of prosody in first language acquisition*, Amsterdam: John Benjamins Publishing Company.

Evans, N. and Levinson, S.C. (2009), The myth of language universals: Language diversity and its importance for cognitive science, *Behavioral and Brain Sciences*, 32 (5): 429–48.

Fagan, M.B. (2012), The joint account of mechanistic explanation, *Philosophy of Science*, 79 (4): 448–72.

Fedurek, P., Zuberbühler, K., and Dahl, C.D. (2016), Sequential information in a great ape utterance, *Scientific Reports*, 6 (1): 38226.

Fitch, W.T. (2020), Animal cognition and the evolution of human language: Why we cannot focus solely on communication, *Philosophical Transactions of the Royal Society B*, 375 (1789): 20190046.

Fodor, J. (1987), *Psychosemantics*, Cambridge: The MIT Press.

Forbes, C.E., Schmader, T. and Allen, J.J., (2008), The role of devaluing and discounting in performance monitoring: A neurophysiologicalstudy of minori ties under threat, *Social Cognitive and Affective Neuroscience*,3 (3): 253-261.

Forbes, C.E. and Schmader, T. (2010), Retraining attitudes and stereotypes to affect motivation and cognitive capacity under stereotype threat, *Journal of Personality and Social Psychology*, 99 (5): 740.

Forbes, C.E. and Leitner, J.B. (2014), Stereotype threat engenders neural attentional bias toward negative feedback to undermine performance, *Biological Psychology*, 102: 98–107.

Freeman, W. (2000), *How brains make up their minds*, New York: Columbia University Press.

Freeman, W. and Skarda, C. (1990), Representations: Who needs them?, in J.L. McGaugh, N.M. Weinberger, and G. Lynch (eds.), *Brain organization and memory: Cells, systems and circuits*, 375–80, New York: Guilford Press.

Friedrich, A., Flunger, B., Nagengast, B., Jonkmann, K., and Trautwein, U. (2015), Pygmalion effects in the classroom: Teacher expectancy effects on students math achievement, *Contemporary Educational Psychology*, 41: 1–12.

Friston, K. (2008), Hierarchical models in the brain, *PLoS Computational Biology*, 4 (11): e1000211.

Froese, T. and Ziemke, T. (2009), Enactive artificial intelligence: Investigating the systemic organization of life and mind, *Artificial Intelligence*, 173 (3–4): 466–500.

Galke, L., Ram, Y., and Raviv, L. (2022), Emergent communication for understanding human language evolution: What's missing?, *Emergent Communication Workshop at ICLR 2022*.

Galke, L., Ram, Y., and Raviv, L. (2023), What makes a language easy to deep-learn?, *arXiv preprint*: 2302.12239.

Gallagher, S. (2017), *Enactivist interventions: Rethinking the mind*. Oxford: Oxford University Press.

Gallagher, S. (2023), Surprise! Why enactivism and predictive processing are parting ways: The case of improvisation, *Possibility Studies & Society*, 1 (3): 269–78.

Gallagher, S. and Zahavi, D. (2020), *The phenomenological mind*. London and New York: Routledge.

Gallagher, S., Hutto, D.D., Slaby, J., and Cole, J. (2013), The brain as part of an enactive system, *Behavioral and Brain Sciences*, 36 (4): 421–2.

Gánti, Tibor. (2003), *The Principles of life*, New York: Oxford University Press.

Garfield, J.L. (2016), Illusion and givenness, *Journal of Consciousness Studies*, 23 (11–12): 73–82.

Glennan, S. (2010), Mechanisms, causes, and the layered model of the world, *Philosophy and Phenomenological Research*, 81 (2): 362–81.

Grice, H.P. (1957), Meaning, *The Philosophical Review*, 66 (3): 377–88.

Griffin, D.R. (1976), *The question of animal awareness: Evolutionary continuity of mental experience*, New York: Rockefeller University Press.

Grinde, B. (2013), The evolutionary rationale for consciousness, *Biological Theory*, 7: 227–36.

Grüter, C. and Czaczkes, T.J. (2019), Communication in social insects and how it is shaped by individual experience, *Animal Behaviour*, 151: 207–15.

Haak, J., Downer, J., and Reeve, R. (2012), Home literacy exposure and early language and literacy skills in children who struggle with behavior and attention problems, *Early Education & Development,* 23 (5): 728–47.

Halina, M., Rossano, F., and Tomasello, M. (2013), The ontogenetic ritualization of bonobo gestures, *Animal Cognition*, 16: 653–66.

Haugeland, J. (2000), *Having thought*. Cambridge: Harvard University Press.

Haugeland, J. (2012), Semantic engines: An introduction to mind design, in *Machine Intelligence,* 29–63, Routledge.

Hauser, M., Chomsky, N., and Fitch, W. (2002), The faculty of language: What is it, who has it, and how did it evolve?, *Science,* 298 (5598): 1569–79.

Heidegger, M. (1962), *Being and time*, J. Macquarrie and E. Robinson (trans). New York: Harper & Row, originally published 1927.

Heidegger, M. (1982), *The basic problems of phenomenology*, A. Hofstadter (trans). Bloomington: Indiana University Press, originally published 1919/20.

Hill, C.S. (2016), Insects: Still looking like zombies, *Animal Sentience,* 9 (20).

Hinton, G.E. (2007), Learning multiple layers of representation, *Trends in cognitive sciences,* 11 (10): 428–434.

Hockett, C. (1977), *The view from language: Selected essays*, Athens: University of Georgia Press.

Hölldobler, B., (1999), Multimodal signals in ant communication, *Journal of Comparative Physiology*, 184 (2): 129–41.

Hommel, B. Müsseler, J., Aschersleben, G., and Prinz, W. (2001), The Theory of Event Coding (TEC): A framework for perception and action planning, *Behavioral and Brain Sciences*, 24: 849–937.

Horwitz, E. (2001), Language anxiety and achievement, *Annual Review of Applied Linguistics*, 21: 112–26.

Hu, T. (2023), Genetic programming for interpretable and explainable machine learning, *Genetic Programming Theory and Practice XIX*, Chapter 4: 81–90.

Huber, M.E., Seitchik, A.E., Brown, A.J., Sternad, D., and Harkins, S.G. (2015), The effect of stereotype threat on performance of a rhythmic motor skill, *Journal of Experimental Psychology: Human Perception and Performance*, 41 (2): 525.

Hurley, S.L. (1998), *Consciousness in action*. Cambridge: Harvard University Press.

Hutto, D.D. and Myin, E. (2012), *Radicalizing enactivism: Basic minds without content*. Cambridge: The MIT press.

Hutto, D.D. and Myin, E. (2017), *Evolving Enactivism*, Cambridge: The MIT Press.

Inzlicht, M., McKay, L., and Aronson, J. (2006), Stigma as ego depletion: How being the target of prejudice affects self-control, *Psychological Science*, 17 (3): 262–69.

Jackson, D.E. and Ratnieks, F.L. (2006), Communication in ants, *Current Biology*, 16 (15): R570–4.

Jeffares, B. (2010), The co-evolution of tools and minds: Cognition and material culture in the hominin lineage, *Phenomenological Cognitive Science*, 9: 503–20.

Jensen, C.N., Burleson, W., and Sadauskas, J. (2012), Fostering early literacy skills in children's libraries: Opportunities for embodied cognition and tangible technologies, *Proceedings of the 11th international conference on Interaction Design and Children*, 50–9.

John-Henderson, N.A., Rheinschmidt, M.L., and Mendoza-Denton, R. (2015), Cytokine responses and math performance: The role of stereotype threat and anxiety reappraisals, *Journal of Experimental Social Psychology*, 56: 203–6.

Johns, M., Inzlicht, M., and Schmader, T. (2008), Stereotype threat and executive resource depletion: Examining the influence of emotion regulation, *Journal of Experimental Psychology: General*, 137 (4): 691.

Johnston, M. (2004), The obscure object of hallucination, *Philosophical Studies*, 120 (1–3): 113–83.

Kaptchuk, T.J., Friedlander, E., Kelley, J.M., Sanchez, M.N., Kokkotou, E., Singer, J.P., Kowalczykowski, M., Miller, F.G., Kirsch, I., and Lembo, A.J. (2010), Placebos without deception: A randomized controlled trial in irritable bowel syndrome, *PloS One*, 5 (12): e15591.

Kaptchuk, T.J. and Miller, F.G. (2015), Placebo effects in medicine, *The New England Journal of Medicine*, 373 (1): 8–9.

Keijzer, F. (1998), Doing without representations which specify what to do, *Philosophical Psychology*, II (3): 269–302.

Keijzer, F. (2002), Representation in dynamical and embodied cognition, *Cognitive Systems Research*, 3 (3): 275–88.

Keijzer, F., Van Duijn, M., and Lyon, P. (2013), What nervous systems do: Early evolution, input–output, and the skin brain thesis, *Adaptive Behavior*, 21 (2): 67–85.

Kelso, J.S. (1995), *Dynamic patterns: The self-organization of brain and behavior*, Cambridge: The MIT press.

Kelso, J.S. and Engstrom, D.A. (2006), *The complementary nature*, Cambridge: The MIT press.

Kim, J. (1998), *Mind in a physical world: An essay on the mind-body problem and mental causation*. Cambridge: The MIT Press.

Kim, H., Somerville, L.H., Johnstone, T., Polis, S., Alexander, A.L., Shin, L.M., and Whalen, P.J. (2004), Contextual modulation of amygdala responsivity to surprised faces, *Journal of Cognitive Neuroscience*, 16 (10): 1730–45.

Kirby, S. (2002), Learning, bottlenecks and the evolution of recursive syntax, in T. Briscoe (ed.), *Linguistic evolution through language acquisition: Formal and computational models*, 173–204, Cambridge: Cambridge University Press.

Kirby, S., Cornish, H., and Smith, K. (2008), Cumulative cultural evolution in the laboratory: An experimental approach to the origins of structure in human language, *Proceedings of the National Academy of Sciences*, 105 (31): 10681–6.

Kirby, S., Dowman, M., and Griffiths, T.L. (2007), Innateness and culture in the evolution of language, *Proceedings of the National Academy of Sciences*, 104 (12): 5241–5.

Kirby, S., Griffiths, T., and Smith, K. (2014), Iterated learning and the evolution of language, *Current Opinion in Neurobiology*, 28: 108–14.

Kirby, S., Smith, K., and Brighton, H. (2004), From UG to universals: Linguistic adaptation through iterated learning, *Studies in Language*, 28 (3): 587–607.

Kirsch, I. (1997), Specifying nonspecifics: Psychological mechanisms of placebo effects, in A. Harrington (ed.), *The Placebo Effect: An Interdisciplinary Exploration*, 166–86, Cambridge, MA: Harvard University Press.

Klein, C. and Barron, A.B. (2016), Insects have the capacity for subjective experience, *Animal Sentience*, 1 (9).

Kok, P., Brouwer, G.J., van Gerven, M.A., and de Lange, F.P. (2013), Prior expectations bias sensory representations in visual cortex, *Journal of Neuroscience*, 33 (41): 16275–84.

Krendl, A.C., Ambady, N., and Kensinger, E.A. (2015), The dissociable effects of stereotype threat on older adults memory encoding and retrieval, *Journal of Applied Research in Memory and Cognition*, 4 (2): 103–109.

Kriegel, U. (2009), *Subjective Consciousness*. Oxford: Oxford University Press.

Lamont, R.A., Swift, H.J., and Abrams, D. (2015), A review and meta-analysis of age-based stereotype threat: Negative stereotypes, not facts, do the damage, *Psychology and Aging*, 30 (1): 180.

Laurin, R. (2013), Stereotype threat and lift effects in motor task performance: The mediating role of somatic and cognitive anxiety, *The Journal of Social Psychology*, 153 (6): 687–99.

Lemon, O. (2022), Conversational grounding in emergent communication—data and divergence, *Emergent Communication Workshop at ICLR 2022*.

Levy, A. and Bechtel, W. (2013), Abstraction and the organization of mechanisms, *Philosophy of Science*, 80 (2): 241–61.

Lewis, D. (1969), *Convention: A philosophical study*. Reprint 2002. Blackwell.

Locatelli, R. and Wilson, K.A. (2017), Introduction: Perception without representation, *Topoi*, 36: 197–212.

Logan, R. (2007), *The extended mind: The emergence of language, the human mind, and culture*. Toronto: University of Toronto Press.

Long, G.M. and Toppino, T.C. (2004), Enduring interest in perceptual ambiguity: Alternating views of reversible figures, *Psychological Bulletin*, 130 (5): 748.

Ludlow, Edward (15 April 2024), Waymo, Cruise and Zoox Inch Forward Ahead of Tesla Joining Robotaxi Race, Bloomberg.com, (Retrieved May 20, 2024).

Lupyan, G. and Dale, R. (2010), Language structure is partly determined by social structure, *PloS One*, 5 (1): e8559.

Lycan, W. (1996), *Consciousness and experience*. Cambridge: The MIT Press.

Lyon, P. and Keijzer, F.A. (2022), The human stain: Why cognitivism cant tell us what cognition is & what it does, in B. Wallace (ed.), *The mind, the world, and the body: Psychology after cognitivism*, 132–65, Throverton: Imprint Academic.

McNamee, G.D., McLane, J.B., Cooper, P.M., and Kerwin, S.M. (1985), Cognition and affect in early literacy development, *Early Child Development and Care*, 20 (4): 229–44.

Macedonia, J.M. and Evans, C.S. (1993), Variation among mammalian alarm call systems and the problem of meaning in animal signals, *Ethology*, 93: 177–97.

Macinnes, I. and Di Paolo, E. (2006), The advantages of evolving perceptual cues, *Adaptive Behavior*, 14 (2): 147–56.

Markman, A. B. and Dietrich, E. (2000), In defense of representation, *Cognitive Psychology*, 40: 138–71.

Martin, M. (2002), The transparency of experience, *Mind & Language*, 17 (4): 376–425.

Martin, M. (2004), The limits of self-awareness, *Philosophical Studies*, 120 (1–3): 37–89

Matheson, K. and Cole, B.M. (2004), Coping with a threatened group identity: Psychosocial and neuroendocrine responses, *Journal of Experimental Social Psychology*, 40 (6): 777–86.

Maturana, H. and Varela, F. (1980), *Autopoiesis and cognition: The realization of the living* (Vol. 42) in Boston Studies in the Philosophy of science, edited by Robert S. Cohen and Marx W. Wartofs. *Dordrecht: D. Reidel Publishing Company.*

Menary, R. (2007), *Cognitive integration*. Basingstoke: Palgrave Macmillan.

Merleau-Ponty, M. (2012), *Phenomenology of perception*, D. A. Landes (trans). London: Routledge, originally published 1945.

Miller, T. (2023), Explainable AI is dead, long live explainable AI! Hypothesis-driven decision support using evaluative AI, *Proceedings of the 2023 ACM Conference on Fairness, Accountability, and Transparency,* 333–42.

Millikan, R.G. (1984), *Language, thought and other biological objects*. Cambridge: The MIT Press.

Michael, R.B., Garry, M., and Kirsch, I. (2012), Suggestion, cognition, and behavior, *Current Directions in Psychological Science*, 21 (3): 151–6.

Mischel, W. and Ebbesen, E.B. (1970), Attention in delay of gratification, *Journal of Personality and Social Psychology*, 16 (2): 329–37.

Mizunami, M., Unoki, S., Mori, Y. et al. (2009), Roles of octopaminergic and dopaminergic neurons in appetitive and aversive memory recall in an insect, *BMC Biology*, 7: 46.

Montgomery, G.H. and Kirsch, I., (1997), Classical conditioning and the placebo effect, *Pain*, 72 (1): 107-113.

Mora, M.S., Nestoriuc, Y., and Rief, W. (2011), Lessons learned from placebo groups in antidepressant trials, *Philosophical Transactions of the Royal Society B: Biological Sciences*, 366 (1572): 1879–88.

Morais, J. and Kolinsky, R. (2005), Literacy and cognitive change, in M.J. Snowling and C. Hulme (eds.), *The science of reading: A handbook,* 188–203, Malden, MA: Blackwell Publishing.

Neander, K. (2017), *A mark of the mental*, Cambridge: The MIT Press.

Nguyen, A., Yosinski, J., Clune, J. (2015), Deep neural networks are easily fooled: High confidence predictions for unrecognizable images, in *Proceedings of the IEEE conference on computer vision and pattern recognition*, 427–436.

Nitschke, J.B., Dixon, G.E., Sarinopoulos, I., Short, S.J., Cohen, J.D., Smith, E.E., Kosslyn, S.M., Rose, R.M., and Davidson, R.J. (2006), Altering expectancy dampens neural response to aversive taste in primary taste cortex, *Nature Neuroscience*, 9 (3): 435–42.

Noë, A. (2004), *Action in perception*. Cambridge: The MIT Press.

Noë, A. (2009), *Out of our heads: Why you are not your brain and other lessons from the biology of consciousness*, New York: Hill and Wang.

Noë, A. (2010), Vision without representation, in N. Gangopadhyay, M. Madary, and F. Spicer (eds.), *Perception, action, and consciousness: Sensorimotor dynamics and two visual systems*, 245–56, Oxford: Oxford University Press.

Noë, A. (2016), Sensations and situations: A sensorimotor integrationist approach, *Journal of Consciousness Studies*, 23 (5–6): 66–79.

Noë, A. (2021), The enactive approach: A briefer statement, with some remarks on "radical enactivism," *Phenomenology and the Cognitive Sciences*, 20 (5): 957–70.

Ofan, R.H., Rubin, N., and Amodio, D.M. (2014), Situation-based social anxiety enhances the neural processing of faces: Evidence from an intergroup context, *Social Cognitive and Affective Neuroscience*, 9 (8): 1055–61.

O'Regan, K. and Noë, A. (2001), A sensorimotor account of vision and visual consciousness, *Behavioral and Brain Sciences*, 24: 939–1031.

Ossenkopf, M., Luck, K.S., and Mathewson, K.W. (2022), Which language evolves between heterogeneous agents?—Communicating movement instructions with widely different time scopes, *Emergent Communication Workshop at ICLR 2022*.

Palmer, A. (2024), Amazon's Zoox under investigation by NHTSA after two crashes. https://www.cnbc.com/2024/05/13/amazons-zoox-under-investigation-by-nhtsa-after-two-robotaxi-crashes.html. (Retrieved May 20, 2024).

Parloff, R. (2016), Why deep learning is suddenly changing your life, *Fortune*.

Pavlov, I.P. (1927), *Conditioned reflexes: An investigation of the physiological activity of the cerebral cortex*, G. V. Anrep (trans). London: Oxford University Press.

Peirce, C.S. (1958), *Collected papers of C. S. Peirce*, C. Hartshorne, P. Weiss, and A. Burks (eds.), Cambridge: Harvard University Press, originally written 1893–1913.

Penn, D., Holyoak, K., and Povinelli, D. (2008), Darwin's mistake: Explaining the discontinuity between human and nonhuman minds, *Behavioral and Brain Sciences*, 31 (2): 109–30.

Pfeifer, R., Iida, F., and Bongard, J. (2005), New robotics: Design principles for intelligent systems, *Artificial Life*, 11 (1–2): 99–120.

Piccinini, G. and Craver, C. (2011), Integrating psychology and neuroscience: Functional analyses as mechanism sketches, *Synthese*, 183: 283–311.

Piqueret, B. and d'Ettorre, P. (2021), Communication in ant societies, in A.B. Kaufman, J. Call, and J.C. Kaufman (eds.), *The Cambridge handbook of animal cognition*, 36–55, Cambridge: Cambridge University Press.

Porro, C.A., Baraldi, P., Pagnoni, G., Serafini, M., Facchin, P., Maieron, M., and Nichelli, P. (2002), Does anticipation of pain affect cortical nociceptive systems?, *Journal of Neuroscience*, 22 (8): 3206–14.

Port, R.F. and Van Gelder, T., eds. (1995), *Mind as motion: Explorations in the dynamics of cognition*. Cambridge: The MIT press.

Prat, Y. (2019), Animals have no language, and humans are animals too, *Perspectives on Psychological Science*, 14 (5): 885–93.

Price, D.D., Finniss, D.G., and Benedetti, F. (2008), A comprehensive review of the placebo effect: Recent advances and current thought, *Annual Review of Psychology*, 59: 565–90.

Price, D.D., Milling, L.S., Kirsch, I., Duff, A., Montgomery, G.H., and Nicholls, S.S. (1999), An analysis of factors that contribute to the magnitude of placebo analgesia in an experimental paradigm, *Pain*, 83 (2): 147–56.

Price, T., Wadewitz, P., Cheney, D. et al. (2015), Vervets revisited: A quantitative analysis of alarm call structure and context specificity, *Science Report*, 5: 13220.

Putnam, H. (1967), The nature of mental states, in W.H. Capitana and D.D. Merrill (eds.), *Mind, and religion*, 37–46, Pittsburgh: University of Pittsburgh Press.

Racine, M., Tousignant-Laflamme, Y., Kloda, L.A., Dion, D., Dupuis, G., and Choinière, M. (2012), A systematic literature review of 10 years of research on sex/gender and pain perception–part 2: do biopsychosocial factors alter pain sensitivity differently in women and men?, *Pain*, 153 (3): 619–35.

Ramsey, W. (2003), Are receptors representations? *Journal of Experimental and Theoretical Artificial Intelligence*, 15 (2): 125–41.

Ramsey, W. (2007), *Representation reconsidered*. Cambridge: Cambridge University Press.

Raviv, L., de Heer Kloots, M., and Meyer, A. (2021), What makes a language easy to learn? A preregistered study on how systematic structure and community size affect language learnability, *Cognition*, 210: 104620.

Raviv, L., Meyer, A., and Lev-Ari, S. (2019), Larger communities create more systematic languages, *Proceedings of the Royal Society B*, 286 (1907): 20191262.

Ray, G. (2018), Invertebrate sentience: Urgent but understudied, *Wild-Animal Suffering Research*.

Reeder, P.A. (2004), Language learnability and the evolution of word order universals: Insights from artificial grammar learning, Unpublished undergraduate honors thesis, Department of Psychology, Cornell University.

Reis, A., and Castro-Caldas, A. (1997), Illiteracy: A bias for cognitive development, *Journal of the International Neuropsychological Society,* 3: 444–50.

Reis, A., Guerreiro, M., and Castro-Caldas, A. (1994), Influence of educational level of non brain-damaged subjects on visual naming capacities, *Journal of Clinical and Experimental Neuropsychology*, 16: 939–42.

Riccio, M., Cole, S., and Balcetis, E. (2013), Seeing the expected, the desired, and the feared: Influences on perceptual interpretation and directed attention, *Social and Personality Psychology Compass*, 7 (6): 401–14.

Righart, R. and Gelder, B.D. (2008), Recognition of facial expressions is influenced by emotional scene gist, *Cognitive, Affective, & Behavioral Neuroscience*, 8: 264–72.

Robinson, M.E., Riley III, J.L., Myers, C.D., Papas, R.K., Wise, E.A., Waxenberg, L.B. and Fillingim, R.B., (2001), Gender role expectations of pain: relationship to sex differences in pain, *The Journal of Pain*, 2 (5): 251-257.

Robinson, M.E., Gagnon, C.M., Riley III, J.L., and Price, D.D. (2003), Altering gender role expectations: Effects on pain tolerance, pain threshold, and pain ratings, *The Journal of Pain*, 4 (5): 284–8.

Rosenthal, D.M., (1998), Two concepts of consciousness in A. Clark and J. Toribio (eds), *Consciousness and Emotion in Cognitive Science*, Routledge, 1-31.

Rosenthal, D. (2005), *Consciousness and mind*, Clarendon Press.

Rosenthal, R. and Jacobson, L. (1966), Teachers expectancies: Determinants of pupils IQ gains, *Psychological Reports*, 19 (1): 115–18.

Roth, F.P. (1984), Accelerating language learning in young children, *Journal of Child Language*, 11: 89–107.

Rowlands, M. (2006), *Body language*. Cambridge: The MIT Press.

Ruiz-Mirazo, K. and Moreno, A. (2004), Basic autonomy as a fundamental step in the synthesis of life, *Artificial life*, 10 (3): 235–59.

Ryle, G. (1949), *The concept of mind*, New York: Hutchinson House.

SAE International Recommended Practice, Taxonomy and Definitions for Terms Related to Driving Automation Systems for On-Road Motor Vehicles, SAE Standard J3016_202104, Revised April 2021, Issued January 2014. https://doi.org/10.4271/J3016_202104.

Salay, N. (2019), Learning how to represent: An associationist account, *The Journal of Mind and Behavior,* 40 (2): 121–45.

Salay, N. (2021), How reductive analyses of content are confused and how to fix them: A critique of varitel semantics, *The Journal of Mind and Behavior,* 42 (2): 109–38.

Sanford, S.D., Kersh, B.C., Thorn, B.E., Rich, M.A., and Ward, L.C. (2002), Psychosocial mediators of sex differences in pain responsivity, *The Journal of Pain*, 3 (1): 58–64.

Savage-Rumbaugh, S. (1990), Language acquisition in a nonhuman species: Implications for the innateness debate, *Developmental Psychobiology*, 23 (7): 599–620.

Sayigh, L.S., El Haddad, N., Tyack, P.L., Janik, V.M., Wells, R.S., and Jensen, F.H. (2023), Bottlenose dolphin mothers modify signature whistles in the presence of their own calves, *Proceedings of the National Academy of Sciences*, 120 (27): e2300262120.

Schel, A., Townsend, S., Machanda, Z., Zuberbühler, K., and Slocombe, K. (2013), Chimpanzee alarm call production meets key criteria for intentionality, *PLoS One* 8 (10): e76674.

Schmader, T. and Johns, M. (2003), Converging evidence that stereotype threat reduces working memory capacity, *Journal of Personality and Social Psychology*, 85 (3): 440–52.

Schmader, T., Johns, M., and Forbes, C. (2008), An integrated process model of stereotype threat effects on performance, *Psychological Review*, 115 (2): 336–56.

Schmicking, D. and Gallagher, S., eds. (2010), *Handbook of phenomenology and cognitive science*. Berlin: Springer Science.

Schönrogge, K., Barbero, F., Casacci, L.P., Settele, J., and Thomas, J.A. (2017), Acoustic communication within ant societies and its mimicry by mutualistic and socially parasitic myrmecophiles. *Animal Behaviour*, 134: 249–56.

Schwarz, K. (2015), *Between pain and math: How expectations shape cognitive processes from neural activity to behaviour*, Berlin: Logos Verlag GmbH.

Searcy, W.A. (2019), Animal communication, cognition, and the evolution of language, *Animal Behaviour*, 151: 203–5.

Sedivy, J.C., Tanenhaus, M.K., Chambers, C.G., and Carlson, G.N. (1999), Achieving incremental semantic interpretation through contextual representation, *Cognition*, 71 (2): 109–47.

Sellars, W. (1956), Empiricism and the philosophy of mind, *Minnesota Studies in the Philosophy of Science*, 1: 253-329.

Sénéchal, M. and LeFevre, J. (2002), Parental involvement in the development of Childrens reading skill: A five-year longitudinal study, *Child Development*, 73 (2), 445–60.

Seyfarth, R. M., Cheney, D. L., and Marler, P. (1980), Monkey responses to three different alarm calls: Evidence of predator classification and semantic communication, *Science*, 210: 801–3.

Shah, D., Powers, J., Tilton, L., Kriegman, S., Bongard, J., and Kramer-Bottiglio, R. (2021), A soft robot that adapts to environments through shape change, *Nature Machine Intelligence*, 3 (1): 51–9.

Shakespeare, W. (1564–1616), *The tragedy of Hamlet, Prince of Denmark*.

Shea, N., (2018), *Representation in cognitive science,* Oxford: Oxford University Press.

Silberstein, M. (2002), Reduction, emergence and explanation, in P. Machamer and M. Silberstein (eds.), *The Blackwell Guide to the Philosophy of Science*, 80–107, Oxford: Black-well.

Slocombe, K., Lahiff, N., Wilke, C., and Townsend, S. (2022), Chimpanzee vocal communication: What we know from the wild, *Current Opinion in Behavioral Sciences*, 46: 101171.

Steele, C.M. and Aronson, J. (1995), Stereotype threat and the intellectual test performance of African Americans, *Journal of Personality and Social Psychology*, 69 (5): 797–811.

Sterelny, K. (2010), Minds: Extended or scaffolded? *Phenomenology and the Cognitive Sciences*, 9 (4): 465–81.

Stewart, J., Gapenne, O., and Di Paolo, E.A. eds. (2010), *Enaction: Toward a new paradigm for cognitive science*. Cambridge: The MIT press.

Stoet, G. and Geary, D.C. (2012), Can stereotype threat explain the gender gap in mathematics performance and achievement?, *Review of General Psychology*, 16 (1): 93–102.

Stone, J., Lynch, C.I., Sjomeling, M., and Darley, J.M. (1999), Stereotype threat effects on Black and White athletic performance, *Journal of Personality and Social Psychology*, 77 (6): 1213.

Struhsaker, T.T. (1967), in S.A. Altmann (ed.), *Social Communication among Primates*, 281–324, Chicago and London: University of Chicago Press.

Suskind, D. (2015), *Thirty million words: Building a childs brain*. New York: Dutton.

Sutton, J., Harris, C.B., Keil, P.G., and Barnier, A.J. (2010), The psychology of memory, extended cognition, and socially distributed remembering, *Phenomenology and the Cognitive Sciences,* 9 (4): 521–60.

Swanson, H., Orosco, M., and Lussier, C. (2015), Growth in literacy, cognition, and working memory in English language learners, *Journal of Experimental Child Psychology,* 132: 155–88.

Thelen, E. and Smith, L.B. (1994), *A dynamic systems approach to the development of cognition and action*. Cambridge: The MIT press.

Thelen, E., Schöner, G., Scheier, C., and Smith, L.B. (2001), The dynamics of embodiment: A field theory of infant perseverative reaching, *Behavioral and Brain Sciences*, 24 (1): 1–34.

Thompson, E. (2007), *Mind in life*. Cambridge: Harvard University Press.

Thompson, R.K., Oden, D.L., and Boysen, S.T. (1997), Language-naive chimpanzees (Pan troglodytes) judge relations between relations in a conceptual matching-to-sample task, *Journal of Experimental Psychology: Animal Behavior Processes*, 23 (1): 31.

Tomasello, M. (2005), *Constructing a language: A usage-based theory of language acquisition*, Cambridge: Harvard University Press.

Tracey, I. (2010), Getting the pain you expect: Mechanisms of placebo, nocebo and reappraisal effects in humans, *Nature Medicine*, 16 (11): 1277–83.

Travis, C. (2013), *Perception: Essays after Frege*. Oxford: Oxford University Press.

Ueda, R., Ishii, T., Washio, K., and Miyao, Y. (2022), Categorial grammar induction as a compositionality measure for emergent languages in signaling games, *Emergent Communication Workshop at ICLR 2022*.

Van Gelder, T. (1995), What might cognition be, if not computation? *The Journal of Philosophy*, 92 (7): 345–81.

Van Gulick, R. (2004), Higher order global states HOGS: An alternative higher order model of consciousness, in R. Gennaro (ed.), *Higher order theories of consciousness*, Amsterdam and Philadelphia: John Benjamins.

Van Leeuwen, C., Verstijnen, I., and Hekkeit, P. (1999), Common unconscious dynamics underlie uncommon conscious effects: A case study in the iterative nature of perception and creation, in J.S. Jordan (ed.), *Modelling consciousness across the disciplines*, Lanhan: University Press of America.

Varela, F.J. (1979), *Principles of biological autonomy*. New York: Elsevier.

Varela, F.J., Thompson, E., and Rosch, E. (1992), *The embodied mind: Cognitive science and human experience*. Cambridge: The MIT press.

Wechsler, M.E., Kelley, J.M., Boyd, I.O., Dutile, S., Marigowda, G., Kirsch, I., Israel, E., and Kaptchuk, T.J. (2011), Active albuterol or placebo, sham acupuncture, or no intervention in asthma, *New England Journal of Medicine*, 365 (2): 119–26.

Weiskopf, D.A. (2011), Models and mechanisms in psychological explanation, *Synthese*, 183: 313–38.

Wells, J.B., Christiansen, M.H., Race, D.S., Acheson, D.J., and MacDonald, M.C. (2009), Experience and sentence processing: Statistical learning and relative clause comprehension, *Cognitive Psychology*, 58: 250–71.

Wheeler, M. (2004), Is language the ultimate artefact? *Language Sciences*, 26: 693–715.

Wheeler, M. (2005), *Reconstructing the cognitive world: The next step*. Cambridge: Bradford Press.

Wheeler, M. (2008), Cognition in context: Phenomenology, situated robotics and the frame problem, *International Journal of Philosophical Studies*, 16 (3): 323–49.

Wheeler, M. (2011), Embodied cognition and the extended mind, in J. Garvey (ed.), *The continuum companion to philosophy of mind*, Continuum, London.

Wheeler, M. (2013), Is cognition embedded or extended? The case of gestures, in Z. Radman (ed.), *The hand, an organ of the mind: What the manual tells the mental*, Cambridge: The MIT Press.

Wheeler, M. (2015), Thinking beyond the brain, in *Educating and building from the standpoint of extended cognition*, 85–104, Lüneburg: Meson Press.

Wheeler, M. (2019), The reappearing tool: Transparency, smart technology, and the extended mind. *AI & society*, 34 (4): 857–66.

Wieser, M.J. and Brosch, T. (2012), Faces in context: A review and systematization of contextual influences on affective face processing, *Frontiers in Psychology*, 3: 471.

Wieser, M.J., Gerdes, A.B., Büngel, I., Schwarz, K.A., Mühlberger, A., and Pauli, P. (2014), Not so harmless anymore: How context impacts the perception and electrocortical processing of neutral faces, *NeuroImage*, 92: 74–82.

Wilcox, K., Roggeveen, A.L., and Grewal, D. (2011), Shall I tell you now or later? Assimilation and contrast in the evaluation of experiential products, *Journal of Consumer Research*, 38 (4): 763–73.

Wittgenstein, L. (2009), P. M. S. Hacker and Joachim Schulte (eds), John Wiley & Sons, originally published 1953.

Wood, R. and Di Paolo, E.A. (2007), New models for old questions: Evolutionary robotics and the 'A not B' error, in F. Almeida e Costa et al. (eds.), *Advances in artificial life: Proc of the 9th Euro. Conf. on artificial life*, 1141–50, Berlin: Springer-Verlag.

Wraga, M., Duncan, L., Jacobs, E.C., Helt, M., and Church, J. (2006), Stereotype susceptibility narrows the gender gap in imagined self-rotation performance, *Psychonomic Bulletin & Review*, 13: 813–19.

Yeong, S., Fletcher, J., and Bayliss, D. (2017), Impact of early home language exposure on phonological and orthographic skills and their contributions to English literacy abilities in English monolingual and Chinese–English bilingual adults, *Applied Psycholinguistics,* 38 (1), 181–210.

Zahavi, D., ed., (2012), *The Oxford handbook of contemporary phenomenology,* Oxford: Oxford Academic.

Zelazo, P. (2015), Executive function: Reflection, iterative reprocessing, complexity, and the developing brain, *Developmental Review,* 38: 55–68.

Zhou, R. and Hu, T. (2023), Evolutionary approaches to explainable machine learning, in *Handbook of evolutionary machine learning*, 487–506, Springer.

Ziemke, T. (2005), Cybernetics and embodied cognition: On the construction of realities in organisms and robots, *Kybernetes*, 34 (1/2): 118–28.

Zubieta, J.K. and Stohler, C.S. (2009), Neurobiological mechanisms of placebo responses, *Annals of the New York Academy of Sciences*, 1156 (1): 198–210.

Index

www.ingramcontent.com/pod-product-compliance
Ingram Content Group UK Ltd.
Pitfield, Milton Keynes, MK11 3LW, UK
UKHW021031041026
15470UKWH00013B/1293